Sixth International Conference on Materials Science and Manufacturing Technology 2024

IOP Conference Series: Materials Science and Engineering
Volume 1316

Coimbatore, India
3-4 May 2024

ISBN: 979-8-3313-0780-6
ISSN: 1757-8981

Printed from e-media with permission by:

Curran Associates, Inc.
57 Morehouse Lane
Red Hook, NY 12571

Some format issues inherent in the e-media version may also appear in this print version.

Institute of Physics
Dirac House, Temple Back
Bristol BS1 6BE UK

Phone: 44 1 17 929 7481
Fax: 44 1 17 920 0979

techtracking@iop.org

Additional copies of this publication are available from:

Curran Associates, Inc.
57 Morehouse Lane
Red Hook, NY 12571 USA
Phone: 845-758-0400
Fax: 845-758-2633
Email: curran@proceedings.com
Web: www.proceedings.com

TABLE OF CONTENTS

IOP Conf. Series: Materials Science and Engineering **1316** (2024) 011001 doi:10.1088/1757-899X/1316/1/011001

Preface

Materials and technologies are significant elements for all kinds of high-tech industries that pave the road for advancements in the manufacturing area. With the rapid development of computer technology, communications technology, and network technology, the traditional manufacturing process has evolved to intelligent manufacturing that is more technologically flexible and efficient.

The Sixth International Conference on Materials Science and Manufacturing Technology 2024 (ICMSMT 2024) was a premier interdisciplinary platform for researchers, academicians, and practitioners worldwide to present recent developments in materials sciences and manufacturing technology. The fifth edition of the ICMSMT conference series (ICMSMT 2024) has been organized with partnered by Akshaya College of Engineering and Technology, Coimbatore, and Diligentec Solutions, Coimbatore, India. The conference received submissions from 08 countries, and the editorial board of ICMSMT has accepted the papers after the stringent screening and review process. ICMSMT is a scientific event and aimed to be conducted in the second week of April every year.

ICMSMT 2024 was organized in an hybrid model with both online and physical participations. Each presenter was given ten minutes for the presentation and three minutes for queries and discussion. The online meetings have been conducted using the Google Meet platform.

Dr. Ramya Muthusamy
Editor / Chair – TPC
ICMSMT 2024

IOP Conf. Series: Materials Science and Engineering **1316** (2024) 011001 doi:10.1088/1757-899X/1316/1/011001

Organizing Committee

Conference Chair(s)
Dr. Thangaprakash Sengodan, Diligentec Solutions, Coimbatore, India
Dr. Nagaraja S, Akshaya College of Engineering and Technology, Coimbatore, India

General Chair(s)
Dr. Baharudin Bin Ismail, Universiti Malaysia Perlis, Malaysia
Dr. Kapilan S, Akshaya College of Engineering and Technology, India

Organizing Chair(s)
Dr. Senthilkumar S, Diligentec Solutions, Coimbatore, India
Dr. Karthik A, Akshaya College of Engineering and Technology, India

Editorial Committee

Editors
Dr. Ramya Muthusamy, Diligentec Solutions, Coimbatore, India
Prof. Selvaraju M, Akshaya College of Engineering and Technology, India

Review Committee

Chair
Dr. Markas Law Lee Lam, National University of Singapore, Singapore

Members
Prof. Sevugan Rajesh, Akshaya College of Engineering and Technology, India
Dr. Sumathi K, Akshaya College of Engineering and Technology, India

Information Contact
Dr. Rajkumar Ramanujam, Akshaya College of Engineering and Technology, India

Publicity and Promotions
Mr. Sivakumar K, Diligentec Solutions, Coimbatore

Technical Program Committee Chairs
Dr. SivaSankari K, Principal, Akshaya College of Engineering and Technology, India
Dr. Sabareesaan K J, Mechanical and Industrial Section, Nizwa College of Technology, Oman
Dr. Moorthi, S R Engineering College, Warangal, India
Dr. Rajamani R, PSG College of Technology, Coimbatore, India

Technical Program Committee

Dr. Jaroslaw Drelich, Michigan Technological University, USA
Dr. Chong Wen Tong, University of Malaya, Malaysia
Dr. Mohamed Arezki Mellal, M'Hamed Bougara University, Algeria
Dr. Khairel Rafezi Ahmad, University Malaysia Perlis, Malaysia
Dr. Mahanthi S D, Michigan State University, USA
Dr. Mohammed Henini, University of Nottingham, UK
Dr. Sharifah Shahnaz Syed Bakar, University Malaysia Perlis, Malaysia
Dr. Alexandru Bogdan Andreescu, Transilvania University of Brasov, Romania

IOP Conf. Series: Materials Science and Engineering **1316** (2024) 011001 doi:10.1088/1757-899X/1316/1/011001

Dr. Norinsan Kamil Othman, University Science Malaysia, Malaysia
Dr. Mitsuru Itoh, Tokiyo Institute of Technology, Japan
Dr. Sanjeev Kumar, Aalborg University, Denmark
Dr. Ramani Kannan, University Technology Petronas, Malaysia
Dr. Satyabrata Das, Indian Institute of Technology, Kanpur
Dr. Singh S P, Indian Institute of Technology, Delhi
Dr. Rajiv O. Dusane, Indian Institute of Technology, Bombay
Dr. Kamal K Kar, Indian Institute of Technology, Kanpur
Dr. Ashish Kumar Mishra, Indian Institute of Technology, Varanasi
Dr. Anand K, Indian Institute of Technology Madras, India
Dr. Sushil Mishra, Indian Institute of Technology, Bombay
Dr. Ajit Panigrahi, CSIR-Institute of Minerals and Materials Technology (IMMT), Bhubaneswar
Dr. Somnath Bhowmick, Indian Institute of Technology, Kanpur
Dr. Patil P, Indian Institute of Technology, Varanasi
Dr. Amrita Bhattacharya, Indian Institute of Technology, Bombay
Dr. Akhilesh Kumar Singh, Indian Institute of Technology, Varanasi
Dr. Krishnakant, Indian Institute of Technology, Delhi
Dr. Alankar Alankar, Indian Institute of Technology, Bombay
Dr. Vanitha C, National Institute of Technology, Warangal
Dr. Tanmoy Maiti, Indian Institute of Technology, Kanpur
Dr. Jitendra Kumar, Indian Institute of Technology, Varanasi
Dr. Rer. Nat. Ravikumar, Indian Institute of Technology, Madras
Dr. Bikash Kumar, CSIR-IMMT, Bhubaneswar
Dr. Ramkumar J, Indian Institute of Technology, Kanpur
Dr. Jyoti Prasad Borah, National Institute of Technology, Nagaland
Dr. Chandana Rath, Indian Institute of Technology, Varanasi
Dr. Mani V N, C-MET, DIETY, Government of India, India
Dr. Bimal Prasad Singh, CSIR-IMMT, Bhubaneswar
Dr. Vivek Verma, Indian Institute of Technology, Kanpur
Dr. Parag Tandaiya, Indian Institute of Technology, Bombay
Dr. Honey John, Cochin University of Science and Technology, Cochin
Dr. Manas K Dalai, CSIR-IMMT, Bhubaneswar
Dr. Nagaraja S, Akshaya College of Engineering Technology, India
Dr. Mathan Kumar N, Akshaya College of Engineering and Technology, India
Prof. Selvaraju M, Akshaya College of Engineering and Technology, India
Dr. Barkunan S R, Akshaya College of Engineering and Technology, India
Dr. Balamurugan, Amrita School of Engineering, Coimbatore, India
Dr. Josephine R L, National Institute of Technology Trichy (NITT), India
Dr. Kulkarni S M, National Institute of Technology (NITK), Surathkal, India
Dr. Gururaj Punekar, National Institute of Technology (NITK), Surathkal, India

IOP Conf. Series: Materials Science and Engineering **1316** (2024) 011002 doi:10.1088/1757-899X/1316/1/011002

Peer Review Statement

All papers published in this volume have been reviewed through processes administered by the Editors. Reviews were conducted by expert referees to the professional and scientific standards expected of a proceedings journal published by IOP Publishing.

- **Type of peer review:** Single Anonymous
- **Conference submission management system:** Morressier
- **Number of submissions received:** 82
- **Number of submissions sent for review:** 52
- **Number of submissions accepted:** 29
- **Acceptance Rate (Submissions Accepted / Submissions Received × 100):** 35.4
- **Average number of reviews per paper:** 2.24
- **Total number of reviewers involved:** 56
- **Contact person for queries:**
  Name: Dr. Ramya Muthusamy
  Email: ramsthangam@gmail.com
  Affiliation: Sengunthar Engineering College

IOP Conf. Series: Materials Science and Engineering **1316** (2024) 012001 doi:10.1088/1757-899X/1316/1/012001

Design and Fabrication of an Electrical Sheet by Waste Wood Fibre Reinforced Functionalized Polypropylene Composite

Nikhil Chauhan[1], Sanjay Palsule[1,2]

[1] Department of Design, Indian Institute of Technology Roorkee, Roorkee, 247667, India
[2] Department of Polymer & Process Engineering, Indian Institute of Technology Roorkee, Saharanpur Campus, Paper Mill Road, Saharanpur, 247001, India

[1]e-mail: n_chauhan@design.iitr.ac.in, [2]e-mail: sanjay.palsule@pe.iitr.ac.in

Abstract. Utilizing natural resources is essential for sustainable development and circular economy. Polymeric composites reinforced with natural fibers are now being used to design commodity, commercial and engineering products. This study develops eco-friendlier composites of waste wood fibre (WWF) reinforced maleic anhydride functionalized (MAF) polypropylene (PP) composite (WWF/MAF-PP) by the Palsule process by extrusion followed by manufacturing by injection moulding. An electric bulb / electrical round sheet has been designed and fabricated using WWF/MAF-PP composites by injection moulding as a replacement for the existing commercial plastic sheet.
Keywords: Maleic anhydride functionalized polypropylene (MAF-PP); waste wood fiber (WWF); Composite; Palsule process; Mechanical properties; bulb / electrical round sheet.

1. Introduction

Owing to the growing interest in environmentally friendly materials and the newly enacted regulations concerning the environment, efforts are being made for the utilization of natural resources to develop eco-friendly and sustainable products. Because of being abundantly available, recyclable, biodegradable, sustainable, and offering good properties, natural fibers have been the subject of significant research for their use as reinforcements for polymer composites. Natural fiber reinforced polymer composites are eco-friendly and sustainable. These composites have been reviewed recently [1].

These composites are being used to design several engineering systems, and the subject has been reviewed [2]. These materials are replacing fully synthetic materials for several applications, including automobiles, furniture, building components, and several commodity products [3-4]. For example, kenaf fiber [5] and plantain fiber [6] reinforced polyethylene composites have been used to design a product for ballistic protection [5], and to design fenders for the auto body [6], respectively. Composites based on polypropylene as polymer and sugarcane bagasse fiber as reinforcement, have been used to design automobile components [7].

This study uses a sustainable waste wood fiber (WWF) as a reinforcement to develop an eco-friendlier composite based on a maleic anhydride functionalized (MAF) polypropylene (MAF-PP) matrix, by the Palsule process, wherein the adhesion between the matrix and the reinforcement is by chemical bonding [8-18]. Palsule process has been used to develop natural fiber reinforced composites based on maleic anhydride functionalized (MAF) - polyethylene, (MAF-HDPE) [8-9], polypropylene (MAF-PP) [10-12], ethylene propylene rubber (MAF-EPR) [13-14], styrene butylene styrene (MAF-

SEBS) [15], acrylonitrile butadiene styrene (MAF-ABS) [16], styrene acrylonitrile (MAF-SAN) [17], and polystyrene (MAF-PS) [18].

2. Materials

A maleic anhydride functionalized polypropylene (MAF-PP) is chosen as a matrix polymer due to its lower cost, easier procurement, widespread availability, and ease of injection molding. It is OPTIM-425, obtained commercially from Pluss Polymers, Gurugram. To choose the best natural fiber, mechanical characteristics and availability of several natural fibers have been examined and contrasted; and to develop an even eco-friendlier composite, waste wood fiber (WWF) is selected as reinforcing natural fiber due to its good holo-cellulose contents and its easy availability, form the discarded wooden products. A discarded wooden product was cut to obtain WWF of an average 3 mm length.

3. Processing

A 15/85 WWF/MAF-PP composite has been developed by the Palsule process using extrusion and injection molding technique. WWF and MAF-PP were dried in a hot air oven and manually mixed in the required proportions. This mixture was then poured into the hopper of twin-screw extruder, having temperatures of 160, 165, 170, 175, 180, 185, and 195°C from the hopper to die. Extruded 15/85 WWF/MAF-PP composite strands were cooled in water, pelletized in a pelletizer, dried, and then injection molded to fabricate their tensile test samples as per the ASTM standards.

Table 1. Combinations of waste wood fiber reinforced maleic anhydride functionalized polypropylene (WWF/MAF-PP) composites (in weight %)

WWF/MAF-PP Composites	Waste wood fiber (WWF) wt. %	Maleic anhydride functionalized polypropylene (MAF-PP) wt.%
0/100	0	100
15/85	15	85

4. Mechanical Testing and Characterization

Tensile testing of the MAF-PP and the developed 15/85 WWF/MAF-PP composite samples was performed on the universaltesting machine as per the ASTM D 638 at a 5 mm/min strain rate.

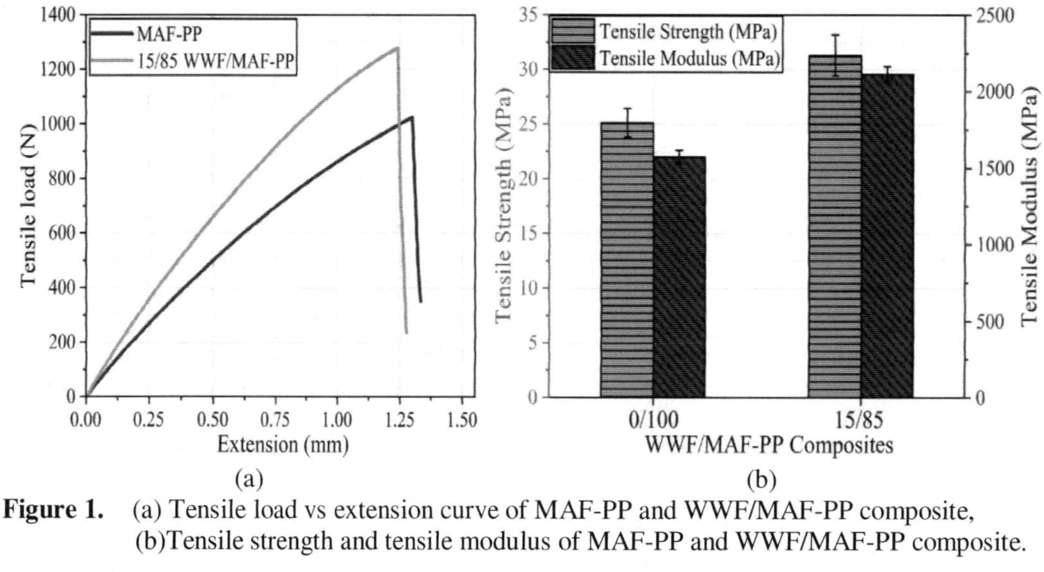

(a) (b)

Figure 1. (a) Tensile load vs extension curve of MAF-PP and WWF/MAF-PP composite, (b)Tensile strength and tensile modulus of MAF-PP and WWF/MAF-PP composite.

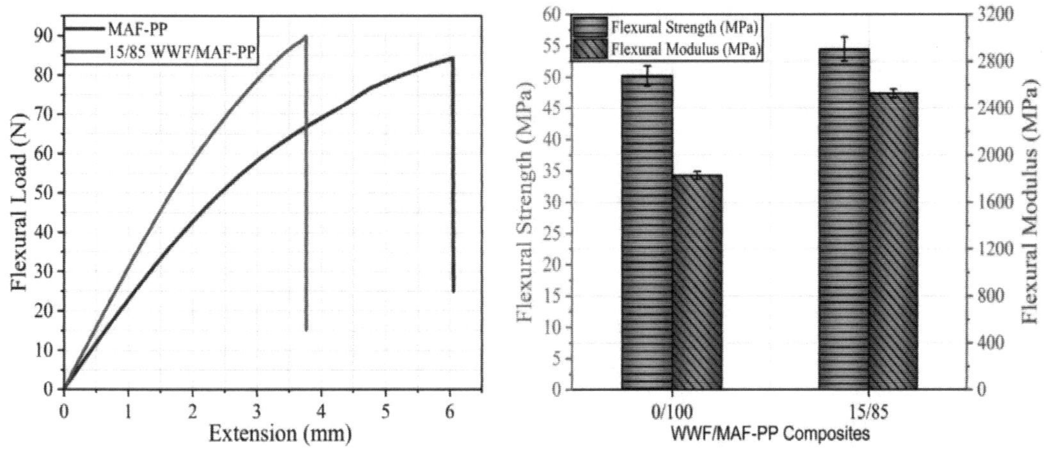

(a) (b)

Figure 2. (a) Flexural load vs extension curve of MAF-PP and WWF/MAF-PP composite, (b) Flexural strength and flexural modulus of MAF-PP and WWF/MAF-PP composite.

The tensile load vs the extension graph of the 15/85 WWF/MAF-PP composite and the MAF-PP matrix are shown in figure1 (a), and their tensile strength and tensile modulus are presented in figure 1 (b). The developed composite has a tensile strength of 31.31 MPa, which is approximately 24.5% higher than that of MAF-PP (25.13 MPa), and a tensile modulus of 2110.99 MPa, which is approximately 34% higher than that of the neat MAF-PP polymer matrix (1574.16 MPa), and its tensile elongation at break is 1.24 mm.

Flexural testing of the generated 15/85 WWF/MAF-PP composite samples and neat MAF-PP was performed as per the ASTM D 790 at a 1.36 mm/min strain rate. The flexural load vs the extension graph of the MAF-PP matrix and the 15/85 WWF/MAF-PP composite are shown in figure2 (a), and their flexural strength and flexural modulus are presented in figure 2 (b). WWF/MAF-PP composite has a flexural strength of approx 54.5 MPa, which is approximately 8.5% higher than that of MAF-PP (50.21 MPa), and a flexural modulus of approx 2529 MPa, which is approximately 38.3% higher than that of the MAF-PP matrix (approx 1828.7 MPa).

5. Design of a Bulb / Electrical Round Sheet

As the first step, following the conceptualization process of identifying workable solutions by evaluating alternatives, through creative thinking, and formulating ideas to generate possible solutions, a conceptual design has been developed. This creation of product design specifications (PDSs) and conceptual and detailed design development was followed by product manufacturing steps as per the standard procedures, referred as [19-21]. The entire idea generation procedure is shown in figure 3.

Bulbs or electrical round sheets are mostly made up of plastic materials like polypropylene, PVC, etc. Of the several available bulb sheet, a commercially available bulb / electrical round sheet obtained from a commercial supplier, which is made up of polypropylene, as shown in figure 4, was selected. Its tensile properties are reported, but flexural are not reported. Its reported tensile strength is 23 MPa, tensile modulus is 1100 MPa. Its diameter is 90 mm, thickness is 1.80 mm and weight is 10 gm.

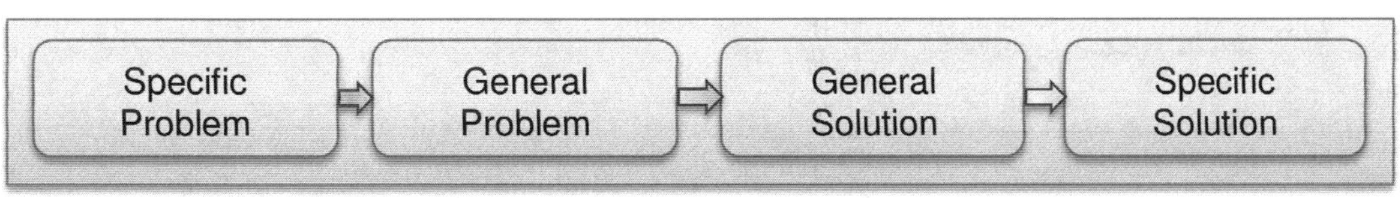

Figure 3. Idea Conceptualization Procedure

IOP Conf. Series: Materials Science and Engineering **1316** (2024) 012001 doi:10.1088/1757-899X/1316/1/012001

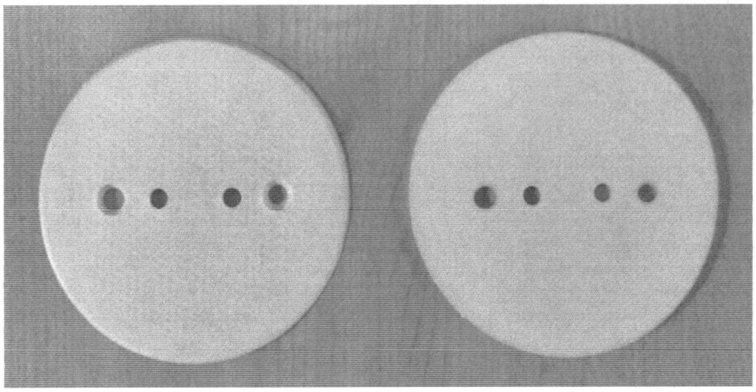

Figure 4. Commercial bulb / electrical round sheet

Design Features	Solutions 1	2	3	4
Fiber type	Synthetic	Natural		
Fiber source	Animal	Mineral	Plant	Leaf
Fiber extraction	Grass	Fruit	Wood	
Fiber name		Hardwood	Softwood	
Polymer	Acrylonitrile butadiene styrene (ABS)	Polypropylene	Polyethylene	Polyvinyl chloride (PVC)
Composite fabrication process	Fiber treatment process	Palsule process	Compatibilizer process	Combined fiber treatment and compatibilizer process
Manufacturing process	Direct-Compression molding	Single-screw extruder	Twin-screw extruder	Direct-Injection molding
Product development process	Rotational molding	Compression molding	Injection molding	

Figure 5. Morphological Chart for Design Approach

This study aims to use the above referred existing bulb or electrical sheet design as a concept and accordingly design and develop a bulb / electrical round sheet using the processed 15/85 WWF/MAF-PP composite. The morphological chart approach shown in figure 5 was also used to produce alternatives. Brainstorming and objective tree methods [22] were then followed to establish objectives, and the process outcomes were utilized to create and produce the design concepts. Potential inputs from brainstorming and morphology chart approach were incorporated into the objective tree diagram, (figure 6), prepared by the following steps: Make a list of objectives for the design, sort the list into groups of objectives that are more and less important, and create a schematic graph of objectives that demonstrates the interconnections and hierarchies between them.

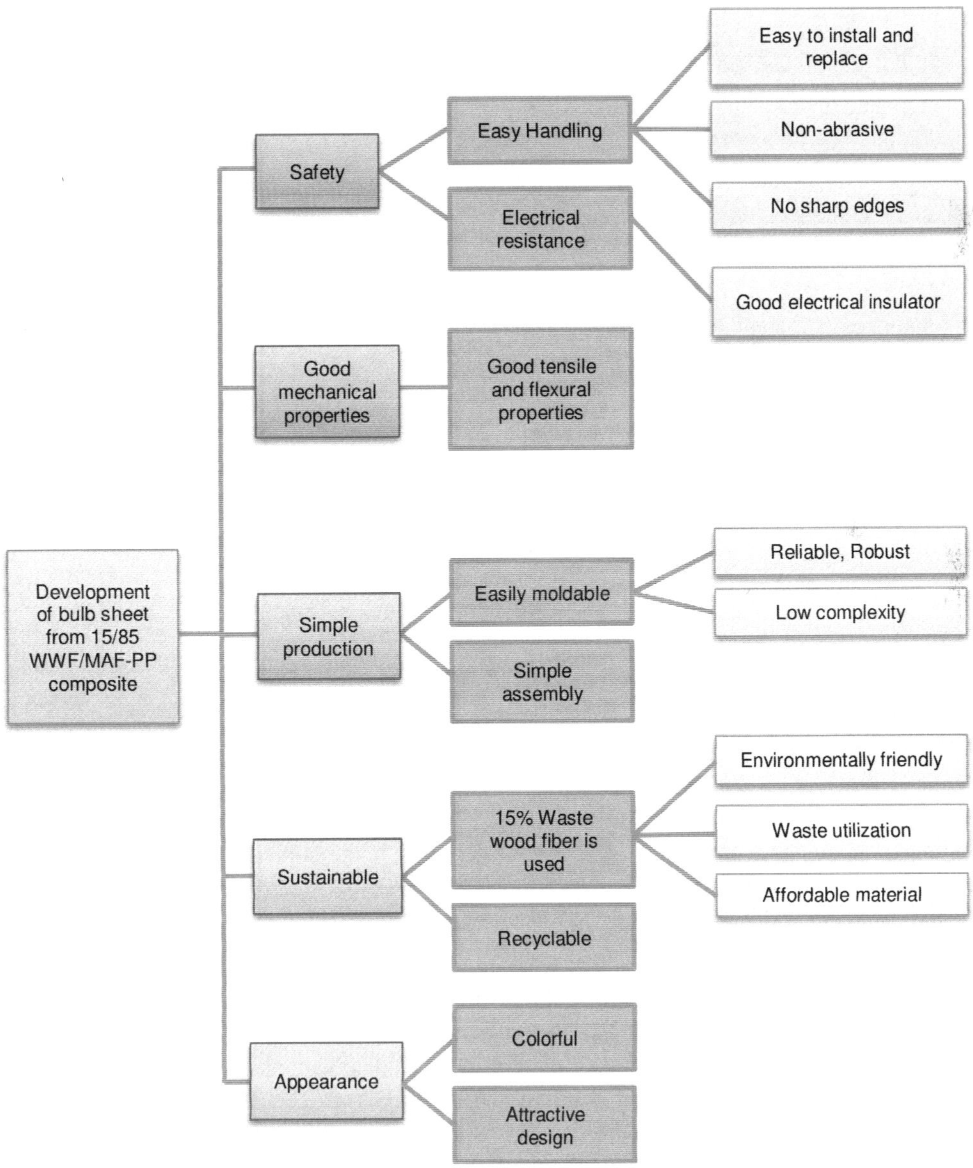

Figure 6. Objective Tree Diagram for Design of the Bulb / Electrical Round Sheet

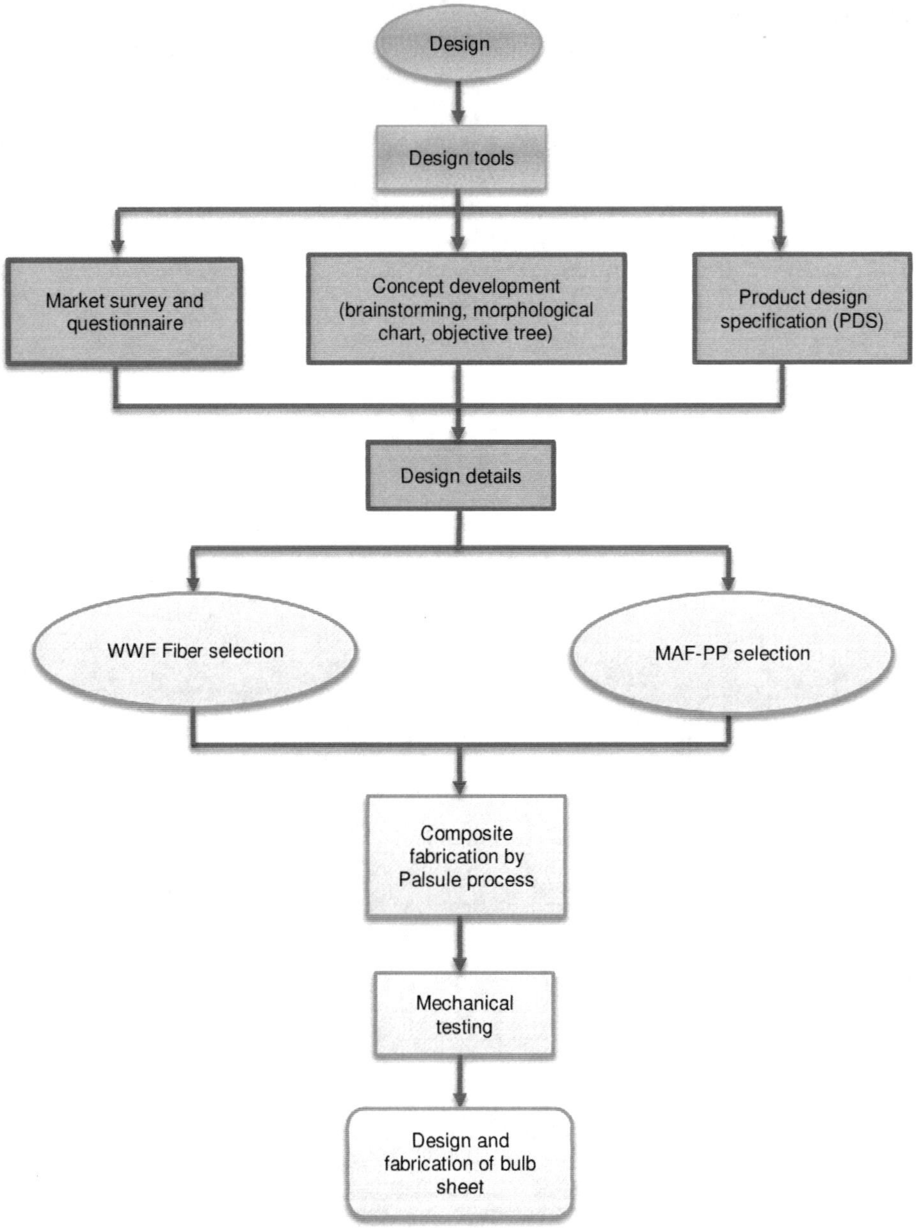

Figure 7. The total design approach for bulb / electrical round sheet composite

6. Manufacturing of Bulb / Electrical Round Sheet

The design inputs developed above were incorporated to fabricate the conceptualized sheet and the total design approach for bulb / electrical round sheet is displayed in figure 7. 15/85 WWF/MAF-PP composite granules, developed by the Palsule process using twin screw extrusion, were firstly dried in an oven at 70°C for 4 hours and then were poured into the hopper of the injection molding machine (Model: Electronica Endura 260). These granules were molded into the shape of bulb / electrical round sheet, by operating the machine with four temperature zones of 185, 190, 195 and 200°C, and with the injection pressure of 65 bar, back pressure of 15 bar, and 15 seconds holding time.

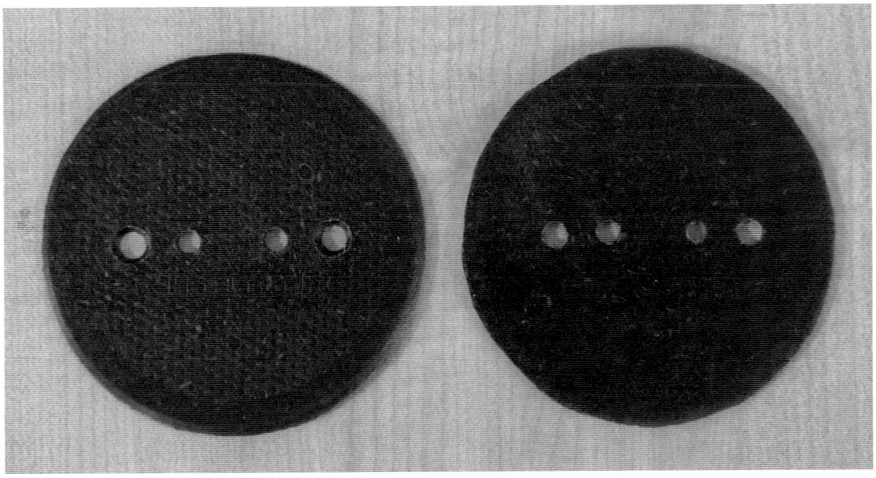

Figure 8. Designed bulb / electrical round sheet and Fabricated Using 15/85 WWF/MAF-PP composite

7. Designed Sheet Fabricated with WWF/MAF-PP Composite

Figure 8 represents the developed bulb / electrical round sheet using 15/85 WWF/MAF-PP composite. This bulb sheet has a 91 mm diameter, 2.04 mm thickness, and 12 gm weight, and the tensile strength of its material is 31.31 MPa, flexural strength is 54.5 MPa, with a tensile modulus of 2110.99 MPa, flexural modulus of 2529 MPa, and elongation at break 1.24 mm. Compared to the commercially available sheet (CS), recorded in section 5, the properties of the fabricated sheet (FS) are: tensile strength approx. 31.31 MPa, (CS 23 MPa), tensile modulus approx. 2.11 GPa, (CS 1.10 GPa), diameter 91 mm (CS 90 mm), thickness 2.04 mm (CS 1.80 mm) and weight 12 gm (CS 10 gm). Thus, the designed and fabricated developed bulb /electrical round sheet (FS) is comparable with commercially available sheet and is more sustainable, being recyclable, eco-friendly, reusable, and harmless. It is non-abrasive, reliable, robust, affordable, attractive and has good mechanical properties.

8. Conclusion

The present study provides a precedent for future research on environmentally friendly sustainable products, employing WWF/MAF-PP composite by examining the mechanical characteristics of waste wood fiber reinforced maleic anhydride functionalized polypropylene composite. Extrusion by Palsule process and manufacturing by injection molding technique is adopted for developing the 15/85 WWF/MAF-PP composite samples. The mechanical characteristics of the molded composite are compared with the mechanical properties of MAF-PP, and the bulb sheet's existing material (PP). The experimental investigations show that the developed 15/85 WWF/MAF-PP composite has better mechanical properties than the reference MAF-PP.

Based on these results, the design, development, and manufacturing of the bulb / electrical round sheet was done using a developed 15/85 WWF/MAF-PP composite. The developed bulb /electrical round sheet is sustainable, recyclable, eco-friendly, reusable, harmless, and has good mechanical properties.

Acknowledgements
Nikhil Chauhan acknowledges the Fellowship received from IIT Roorkee (India).

References

[1] Elfaleh I, Abbassi F, Habibi M, Ahmad F, Guedri M, Nasri M and Garnier C 2023 A comprehensive review of natural fibers and their composites: An eco-friendly alternative to conventional materials. *Results in Engineering*. doi: 10.1016/j.rineng.2023.101271.

[2] Azman M A, Asyraf M R M, Khalina A, Petrů M, Ruzaidi C M, Sapuan S M and Suriani M J 2021 Natural fiber reinforced composite material for product design: A short review. *Polymers*. doi:10.3390/polym13121917.

[3] Prasad S V N B, Kumar G A, Sai K V P and Nagarjuna B 2019 Design and optimization of natural fibre reinforced epoxy composites for automobile application. *AIP Conference Proceedings*. doi: 10.1063/1.5117928.

[4] Kong C, Park H and Lee J 2014 Study on structural design and analysis of flax natural fiber composite tank manufactured by vacuum assisted resin transfer molding. *Mater. Lett.* **130** 21–25.

[5] Akubue P C, Igbokwe P K and Nwabanne J T 2015 Production of Kenaf Fibre Reinforced Polyethylene Composite for Ballistic Protection. *International Journal of Scientific & Engineering Research.* **6** 1–7.

[6] Chukwutoo I C, Obiafudo O J and Okafor C E 2016 Characterization of Plantain Fiber Reinforced High Density Polyethylene Composite for Application in Design of Auto Body Fenders. *Journal of Innovative Research in Engineering Sciences*. **4** 574–587.

[7] Luz S M, Caldeira Pires A and Ferrao P M C 2010 Environmental benefits of substituting talc by sugarcane bagasse fibers as reinforcement in polypropylene composites: Ecodesign and LCA as strategy for automotive components. *Resources, Conservation and Recycling*. **54** 1135–1144.

[8] Singh A A and Palsule S 2013 Jute fiber-reinforced chemically functionalized high-density polyethylene (JF/CF-HDPE) composites with in situ fiber/matrix interfacial adhesion by Palsule Process. *Composite Interfaces*. **20** 553-573.

[9] Singh A A and Palsule S 2014 Coconut fiber reinforced chemically functionalized high-density polyethylene (CNF/CF-HDPE) composites by Palsule process. *Journal of Composite Materials*. **48** 3673-3684.

[10] Priyanka and Palsule S 2013 Banana fiber/chemically functionalized polypropylene composites with in-situ fiber/matrix interfacial adhesion by Palsule process. *Composite Interfaces*. **20** 309–329.

[11] Singh A A and Palsule S 2016 Jute fiber reinforced chemically functionalized polypropylene self-compatibilizing composites by Palsule process. *Journal of Composite Materials*. **50** 1199–1212.

[12] Biswas K and Palsule S 2016 Bamboo fibre-reinforced self-compatibilizing functionalized polypropylene composites by Palsule process. *Polymers and Polymer Composites*. **24** 663-674.

[13] Dinesh and Palsule S 2020 Structure and properties of recycled bamboo fiber reinforced chemically functionalized ethylene propylene rubber composites. *Polymers and Polymer Composites*. **28** 609-622.

[14] Dinesh and Palsule S 2021 Bagasse fiber reinforced functionalized ethylene propylene rubber composites by Palsule process. *Journal of Natural Fibers*. **18** 1637-1649.

[15] Dinesh and Palsule S 2022 Bagasse fiber reinforced functionalized styrene ethylene butylene styrene composites by Palsule process. *Journal of Natural Fibers*. **19** 3981-3992.

[16] Kumar P and Palsule S 2023 Bamboo Fiber Reinforced Chemically Functionalized Acrylonitrile Butadiene Styrene Composites by Palsule Process. *Journal of Natural Fibers* **20** 2150741.

[17] Tamta N and Palsule S 2023 Kenaf fiber reinforced chemically functionalized styrene-acrylonitrile composites. *Journal of Thermoplastic Composite Materials*. doi:10.1177/08927057231168560.

[18] Verma H and Palsule S 2023 Bagasse fiber reinforced chemically functionalized polystyrene composites. *Journal of Thermoplastic Composite Materials*. doi:10.1177/08927057231172665.
[19] Mazani N, Sapuan S M, Sanyang M L, Atiqah A and Ilyas R A 2019 Design and fabrication of a shoe shelf from kenaf fiber reinforced unsaturated polyester composites. *Elsevier Inc*. doi: 10.1016/B978-0-12-816354-2.00017-7.
[20] Wright I C 1997 A review of research into engineering change management: Implications for product design. *Des. Stud.* **18** 33–42.
[21] Pahl G, Beitz W, Feldhusen J and Grote K H 2007 Engineering Design: A Systematic Approach. Third edition.
[22] Cross N 2021 Engineering design methods: strategies for product design. *John Wiley & Sons*. Chichester (England) (UK).

Experimental Research on the effect of Structural glass fibre in the Tensile Strength Ratio and Retained Stability of VG40 Bituminous mix

Aishwarya R*, Hemamathi A, Binu Sukumar, Yashwantha Kumar M, Vijay Bhagavathi P, Arulshankar B

Department of Civil Engineering, R.M.K Engineering College, Tamil Nadu, India

**Corresponding author* raa.civil@rmkec.ac.in

Abstract. The tensile characteristics of bituminous mix are a major concern for pavement engineers due to the cracking issues. The cracking infuses moisture into the pavement layers which affects the stability of the pavement. The measures to reduce the fatigue tensile cracks have to be taken in the design stage rather than the maintenance stage. The tensile properties and the stability are evaluated using the indirect tensile strength and retained stability tests. While fibers improve the tensile properties of pavement, the S-glass fiber possessing superior mechanical strength has been found to have less scope in past studies. Thus, the current study primarily focuses on examining how Structural Glass fiber reinforced in VG-40 bituminous concrete mix improves the tensile and moisture susceptibility characteristics. The Marshall technique of bituminous mix design was used to establish the ideal bitumen content for casting the bituminous concrete specimens. At the ideal bitumen content, the attributes of bituminous mixes were established. Tensile strength ratio (TSR), Indirect Tensile Strength (ITS), and Retained Stability of modified bituminous concrete mix were assessed. It is observed that the bituminous mix prepared by adding the S-Glass fibre of 8mm length by 3% weight of aggregates exhibited higher ITS, TSR, and Retained stability. The bituminous mix with glass fiber incorporated had higher retained stability (95.36%) than the conventional mix (91.17%). The results of the Indirect Tensile Strength Test revealed that the mix with glass fiber added had a higher Tensile Strength Ratio (92%) compared to the mix with conventional fiber (87%). A noteworthy outcome that creates opportunities for more study in this field is the rise of 5% TSR and 4.66% maintained the stability of the glass fiber-modified bituminous mix.

Keywords. Tensile strength ratio, Retained Stability, Marshall Stability, Glass fibre, Bituminous concrete.

1. Introduction

Bituminous pavements are adopted widely all over the world due to the ease of construction and their nature to immediately open for traffic. The structure of the bituminous pavements is flexible in nature due to the grain-to-grain transfer of the vehicular load to the soil subgrade [1]. The performance of the flexible pavements depends highly on the stiffness of the mix and the inter-particle friction available. The cohesion that the paving mix offers to the aggregates and filler materials needs to be sufficient enough to coat the particles as well as not in excess which will keep the pavement materials apart in the presence of excess bitumen. This primary requirement of the bituminous mix is termed 'stability'. In a study [2] conducted on the retained stability of VG30 DBM, the addition of mineral fillers was concluded to have higher retained stability which will reduce the secondary compaction of the pavement due to traffic and resist moisture-prone internal damage of the pavement. The tensile properties of the Bituminous concrete reflect the tensile stresses and strains in the bottom of the bituminous layers due to continuous flexing on the application of vehicular loading. [3] The tensile strength of the paving mix decides its resistance to fatigue cracking, rutting, and adherence of the pavement constituents. [4] The localized and channelized depression of pavements in the vehicular path, rutting, leads to early failure of the road sections which imposes high economic burden to the road maintenance activities. However, unattended fatigue cracking, may also lead to potholes on the roads which imposes a serious threat of moisture infiltration into the layers of the pavements causing stripping failure and the loss of stability. Hence, various types of fibres have been utilized in bituminous concrete to enhance its performance [5]. The application of fibres improves engineering properties, fatigue resistance, rutting resistance and stiffness [6,7]. Numerous research has been conducted to investigate the characteristics and functionality of HMA mixtures incorporating different types of fibre, including basalt, polyester, steel, carbon, aramid, polyethylene, and polypropylene [8,9]. The influence of polymer fibers was reviewed on the mechanical properties of asphalt mixtures using finite element analysis and experimental study [10]. [11] Though glass fibers are used in grid format as a Semi Absorbing Membrane Interlayer (SAMI) in pavements, Glass fibers can provide HMA more stiffness and tensile strength, but handling them during the asphalt manufacturing process is crucial. The use of glass fibers increased the material's resilience to rutting and fatigue cracking. It was mixed and dispersed randomly to successfully prevent internal aggregate dislocation by resisting shear displacement [12]. Glass fibers enhance the fracture energy and crack intensity factor substantially [13]. Asphalt mixtures modified with glass fibers are also more resistant against rutting, crack initiation and propagation [14]. The performance of asphalt mixtures with glass fibers in terms of fracture and rutting was investigated in a different study [15]. The results were consistent with earlier research, and adding glass fibers considerably improved the HMA mixtures' resistance to rutting and fracturing. The Marshall design method was employed in the work provided here to prepare the HMA. Glass fibers (GF) were prepared by adding varying amounts of GF to HMA specimens. The experimental outcomes were contrasted with specification limitations and a control mixture. Consequently, this investigation demonstrated that including GF in an appropriate ratio is a practical means of enhancing the properties and functionality of HMA mixes.

2. Objectives

The primary goal of this study is to ascertain the Optimum Glass Fiber Content for S-Glass Fibre to attain the highest Indirect Tensile Strength and Retained Stability values.

3. Material Characterization

Due to its ease of production and maintenance, bituminous concrete is the most commonly used type of pavement out of all others. It is a composite material, either sandwich or monolithic, that combines the elements of bitumen as a binder with different types of aggregates in predefined quantities [15]. In addition, Glass fibers are used in the study to improve the performance of the bituminous concrete.

3.1 Aggregates

Aggregates exhibit good interlocking properties and offer compressive and shear strength. Tests were performed on aggregates in the laboratory to evaluate their characteristics, such as water absorption, hardness, unit weight, specific gravity, and shape. The combined specific gravity and unit weight of the aggregates were found to be 2.67 and 1429.5 kg/m^3 respectively. The test results are shown in Table 1.

Table 1. Aggregate test results

Properties	Laboratory test	As per MoRTH	Whether within permissible limits?
Aggregate Impact Value Test – IS 2386(Part IV):1978	15.48%	Maximum 24%	Yes
Los Angeles Abrasion Value – IS 2386(Part IV):1978	22.45%	Maximum 30%	Yes
Combined Flakiness and Elongation Index Test–IS 2386(Part I):1978	26.93%	Maximum 35%	Yes
Water Absorption Test – IS 2386 (Part III):1978	0.4%	Maximum 2%	Yes
Stripping Value Test – IS 6241:1971	100%	Minimum Value Recommended as per IRC 111:2009 is 95%	Yes

3.2 Bitumen

In the present study viscosity grade (VG-40) bitumen is used as binder material. In order to check the suitability of bitumen as binder material, tests were conducted in laboratory. The test results satisfy the requirements as per IS 73:2013 and are presented in Table 2. The specific gravity of bitumen (VG-40) is 1.046.

Table 2. Bitumen (VG-40) test results

Properties	Laboratory test	As per IS 73:2013	Whether within permissible limits?
Penetration at 25 °C, 100gm, 5 s, 0.1 mm– IS 1203:1978	37	Minimum 35	Yes
Softening Point (Ring and Ball)– IS 1205:1978	58°C	Minimum 50°C	Yes
Absolute Viscosity (poise) – IS 1206 (Part II):1978	4022.2	3200 – 4800	Yes
Flash Point - IS 1448 [P: 69]	285°C	Minimum 220°C	Yes
Ductility @27 °C, cm	85	Minimum 75	Yes

3.3 Glass Fibre

Structural glass fibre which possesses high tensile and flexural strength is used in this research and it properties are given in Table 3.

Table 3. Properties of Glass fibre

Properties	Detail
Shear Modulus	35 GPa
Tensile Strength	4700 MPa
Young's modulus	86 GPa
Colour	White
Density (g/cm^3)	2.6
Melting point (°C)	>300
Length(mm)	6mm, 8mm, 10mm, 12mm

3.4 Design of Bituminous mix

The aggregate gradation has a significant impact on the hot mix asphalt's permanent deformation. When tested in compliance with IS:2386 Part 1, the total grading of the additional filler and coarse and fine aggregate must stay within the parameters shown in the table. Table 4 shows the cumulative gradation range for the various aggregate sizes as stated in MoRTH. The grain-size distribution curve satisfying the requirements stated by codal provisions is also shown in Figure 1. From the table and the graph, it is evident that 10% of 19mm aggregates, 18% of 13.2mm aggregates, 20% of 6.7mm aggregates and 52% of 2.36mm aggregates is the optimum job mix for the ideal grading pattern according to MoRTH.

Table 4. Gradation of aggregates

Size of Agg.			19.0 mm	13.2 mm	9.50 mm	4.75 mm	2.36 mm	1.18 mm	0.6 mm	0.3 mm	0.15 mm	0.075 mm
19.0 Mm	% Passed		100.00	27.70	1.25	0.00	0.00	0.00	0.00	0.00	0.00	0.00
	Job Mix %	10%	10.00	2.77	0.13	0.00	0.00	0.00	0.00	0.00	0.00	0.00
13.2 mm	% Passed		100.00	98.15	32.15	1.38	0.00	0.00	0.00	0.00	0.00	0.00
	Job Mix %	18%	18.00	17.67	5.79	0.25	0.00	0.00	0.00	0.00	0.00	0.00
6.7 mm	% Passed		100.00	100.00	100.00	47.93	6.98	0.00	0.00	0.00	0.00	0.00
	Job Mix %	20%	20.00	20.00	20.00	9.59	1.40	0.00	0.00	0.00	0.00	0.00
2.36 mm	% Passed		100.00	100.00	100.00	100.00	86.10	74.20	57.60	37.10	24.10	10.2
	Job Mix %	52%	52.00	52.00	52.00	52.00	44.77	38.58	29.95	19.29	12.53	5.30
Combined Grading		**100%**	**100.0**	**92.44**	**77.91**	**61.83**	**46.17**	**38.58**	**29.95**	**19.29**	**12.53**	**5.30**
MORTH Specification			**100**	**90-100**	**70 - 88**	**53 - 71**	**42 - 58**	**34 - 48**	**26 - 38**	**18 - 28**	**12 - 20**	**4 - 10**

4. Experimental work and Results

The Marshall mix design was used to calculate the Optimum Binder Content. The specimens were tested for Marshall Flow Values and graphical plots were sketched. These parameters were used to compute the ideal percentage of Glass Fibre. Finally, the specimens with glass fibre were tested for Indirect Tensile strength test and Retained stability test. Additionally, all of the examined specimens' attributes were contrasted with those of the control mixture that was devoid of GF additions.

4.1 **Marshall Method for Stability and Flow Values for conventional mix**

The ideal amount of binder is determined by the Marshall mix design technique. The performance prediction measure for the Marshall mix design method is provided by the Marshall stability and flow test [16]. The test specimen is loaded at a rate of 50.8 mm/minute and is measured during the stability phase of the test. The specimen is loaded until it fails, at which point the maximum load is deemed stable. A connected dial gauge monitors the plastic flow, or deformation, caused by the loading. At the same time as the maximum load is recorded, the flow value is also recorded in increments of 0.25 mm (0.01 inch).

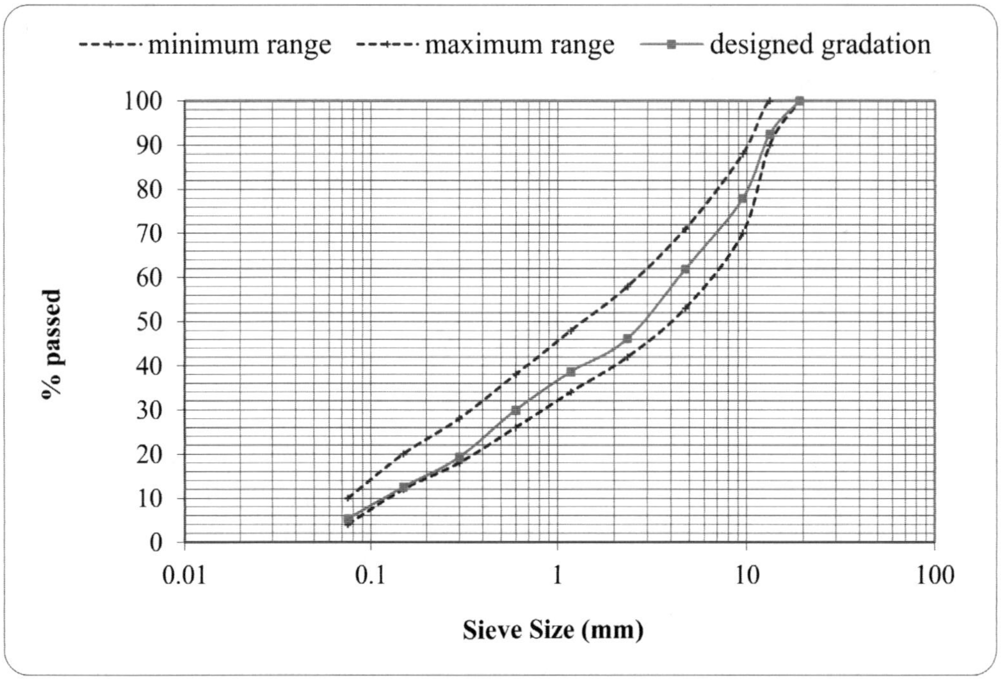

Figure 1. Grain size distribution curve

Table 5. Volumetric analysis and Marshall Stability and Flow Values for Conventional mix

Specimen No.	Qty. of Bitumen Content Added	Wt. of specimen in Air (g)	Wt. of specimen in Water (g)	Weight of Saturated Surface Dry (g)	Stability (kN)	Flow (mm)
1		1257.9	718.1	1258.3	14.3	2.25
2	5%	1258.4	719.5	1260.4	17.3	2.73
3	(60g)	1261.5	716.9	1262.7	15.6	2.80
				Average	**15.733**	**2.593**
4		1263.0	733.1	1263.8	18.2	3.30
5		1267.7	733.5	1268.2	11.7	3.93
6	5.5% (66g)	1264.1	730.5	1265.0	19.7	3.89
				Average	**16.533**	**3.707**
7		1270.0	739.5	1271.3	12.5	3.64
8	6%	1272.3	738.0	1273.6	16	3.88
9	(72g)	1265.4	735.0	1266.8	16.8	3.63
				Average	**16.433**	**3.750**
10		1275.2	742.0	1276.6	15.3	3.90
11	6.5%	1275.1	741.8	1275.9	16.2	4.14
12	(78g)	1278.7	740.5	1279.7	17	4.18
				Average	**16.167**	**4.073**

IOP Conf. Series: Materials Science and Engineering **1316** (2024) 012002 doi:10.1088/1757-899X/1316/1/012002

The variation of Binder Content vs Marshall Stability, Marshall flow, Percentage of Voids (Vv), Voids filled with Bitumen (VFB), and Bulk Specific Gravity [17] as per the graphs in Figure 2- Figure 6 were noted for the conventional mix. As per Marshall Mix Design Specifications, the binder content corresponding to the maximum bulk specific gravity (figure 4) is found to be 6%, Binder content drawn from the graph (figure 5) at (3-5% optimum) 4% air voids was found to be 5.66% and that corresponding to maximum stability (figure 6) is 5.63%. Taking average, the Optimum Binder Content is 5.79% by weight of aggregates for which the flow value from Figure 2 is found as 3.7mm (2-4mm optimum).

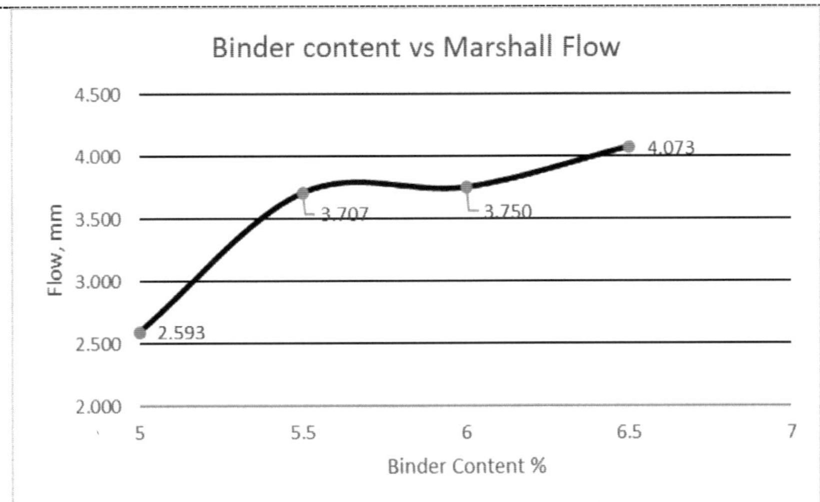

Figure 2. Binder Content vs Marshall Flow

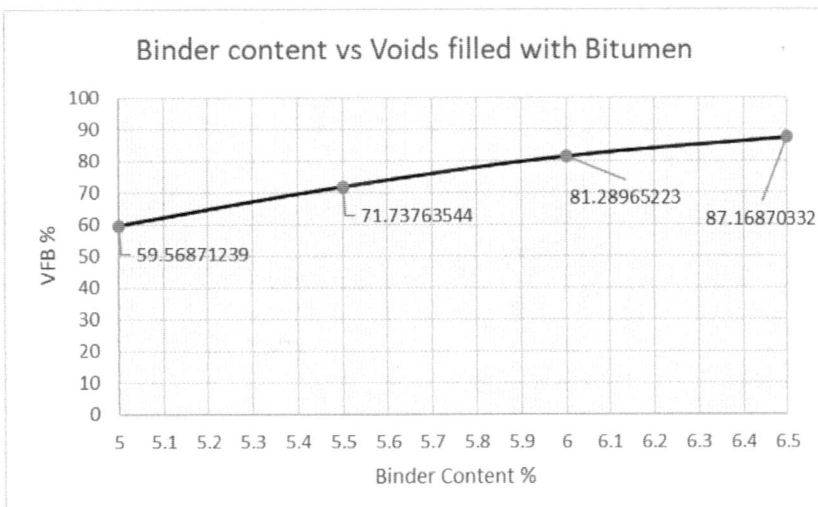

Figure 3. Binder Content vs VFB

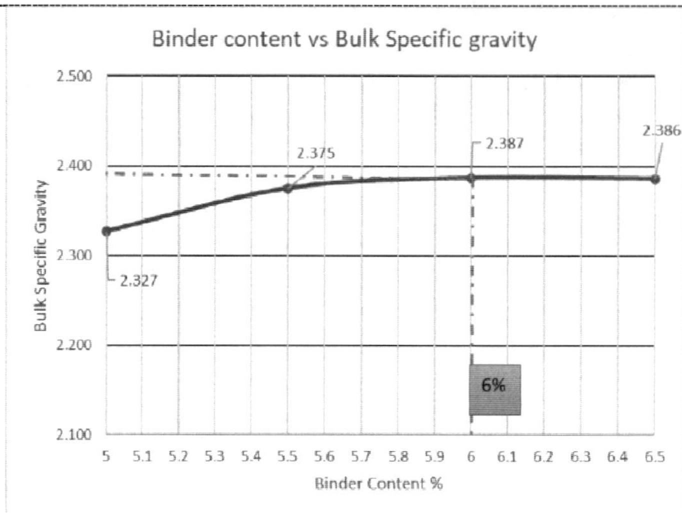

Figure 4. Binder Content vs Bulk Specific Gravity

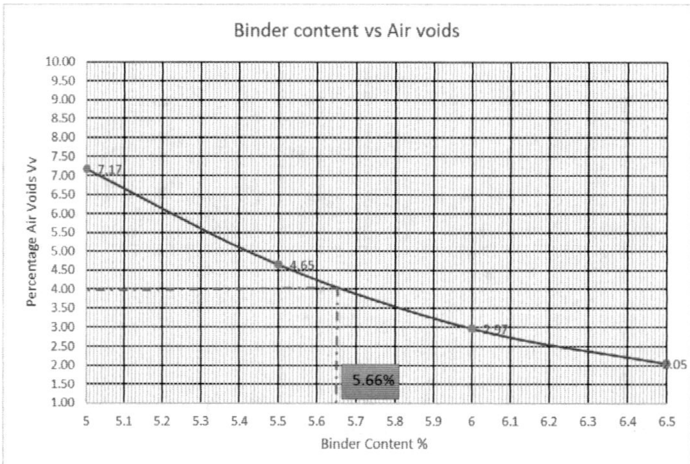

Figure 5. Binder Content vs VFB

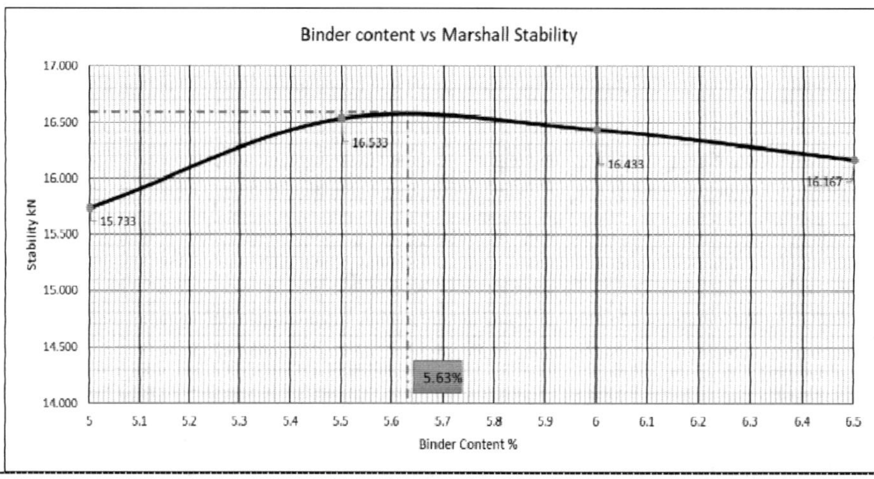

Figure 6. Binder Content vs Marshall Stability

4.2 Glass Fibre Modified Bituminous mix

Bituminous mix with Optimum Binder Content of 5.79% and varying proportions of different lengths of glass fibre were prepared and subjected to Marshall stability test. The test results as given in Table 6.

Table 6. Volumetric analysis and Marshall Stability and Flow Values for Modified mix

Sample No.	Size of Glass Fibre	% of Glass Fibre	Wt. in Air (g)	Wt. in Water(g)	SSD Wt. (g)	Stability (kN)	Flow (mm)
1	6 mm	0.2	1257	742.2	1257.9	19.1	3.44
2			1267	747.5	1268.4	18.8	3.57
3			1263	744.9	1264.3	18.7	4.11
				Average Value =		**18.9**	**3.71**
4		0.3	1268	743.3	1269.8	17	3.61
5			1268	746	1268.6	18.9	4.01
6			1269	735.7	1271.4	12.4	3.93
				Average Value =		**16.1**	**3.85**
7		0.4	1264	743.6	1264.2	24.7	3.96
8			1263	738.2	1264	20	3.65
9			1262	744.1	1264	23.3	4.32
				Average Value =		**22.7**	**3.97**
10	8 mm	0.2	1270	763.1	1271.3	23.7	3.36
11			1266	759.3	1267.4	25.3	3.44
12			1272	759.2	1272.8	19.6	3.59
				Average Value =		**19.2**	**3.46**
13		0.3	1264	759.8	1265	23.7	3.21
14			1266	759.1	1267.8	25.3	3.52
15			1263	758.1	1264.8	19.6	3.13
				Average Value =		**22.9**	**3.28**
16		0.4	1272	760.9	1273.6	22.2	4.03
17			1263	752.7	1264.1	26.2	4.08
18			1266	756.8	1268.1	21.8	4.28
				Average Value =		**23.4**	**4.13**
19	10 mm	0.2	1255.9	752	1256.4	24	3.18
20			1269.4	760.5	1269.8	18.7	3.21
21			1267.7	758	1268.5	18.7	3.26
				Average Value =		**20.5**	**3.22**
22		0.3	1267	754.8	1267.8	17.3	3.4
23			1264.6	755.6	1265.4	22.7	3.56
24			1267.7	750.7	1258.8	20	3.46
				Average Value =		**20**	**3.47**
25		0.4	1259.1	752.3	1260.1	22	4.13
26			1270.3	754.9	1271.5	18.8	4.02
27			1262.2	755.2	1263	24.6	4.21
				Average Value =		**21.8**	**4.12**
28	12 mm	0.2	1268.3	756	1268.8	21	3.23
29			1283.4	766.1	1285.6	21.3	3.21
30			1261.6	755.9	1262.8	21.8	3.36
				Average Value =		**21.4**	**3.26**
31		0.3	1264.5	755.4	1267.8	23.2	3.69
32			1264.5	757.8	1266.6	22.1	3.94
33			1264.6	754.9	1266.2	20.2	3.48
				Average Value =		**21.8**	**3.7**
34		0.4	1262.6	752.9	1264	20.2	3.96
35			1269.8	757.5	1271.4	19.7	4.06
36			1263.8	755.2	1264.8	19.7	4.13
				Average Value =		**19.9**	**4.05**

As per IRC 111:2009, the stability should be greater than 9kN and the flow value should lie between 2-4mm. The highest stability value for flow between 2-4mm is observed for 8mm length GF for 0.3% as shown in figure 3.

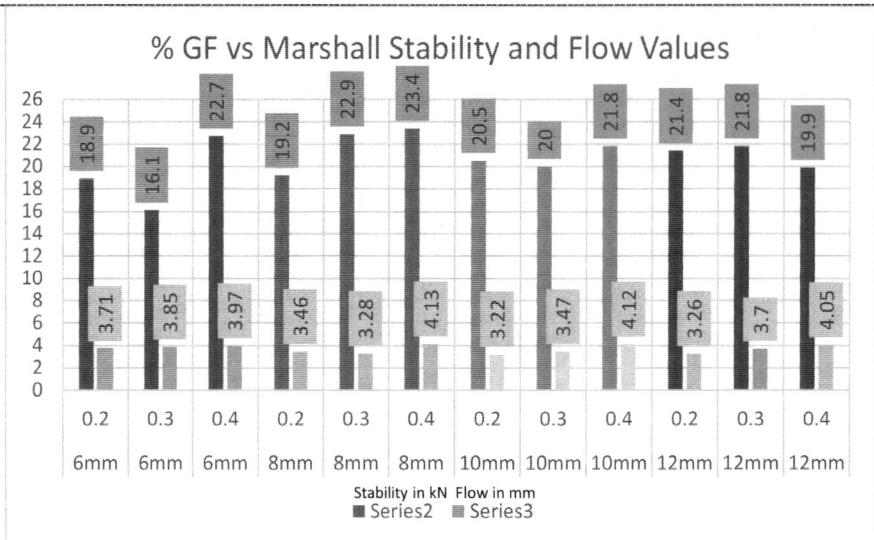

Figure 7. Marshall Stability and Flow values for GF modified mix

4.3 Indirect Tensile Strength Test

[18,19] Long-term traffic loads may have an impact on an asphalt mixture's strength, resulting in fatigue cracks and/or rutting. Indirect Tensile Strength (ITS) is a regularly used metric to assess this damage. The cylindrical specimen must be inserted between the loading strips of the compression testing equipment and loaded diametrically along the cylinder axis at a constant displacement speed until it breaks to assess the ITS. The highest tensile stress is determined by the specimen's dimensions and peak load delivered at the break to determine the indirect tensile strength. The specimen's indirect tensile strength (ITS) was calculated using the formula given below.

$$S_t = \frac{2000P}{\pi t d}$$

Where, S_t = tensile strength, kPa

P = maximum loads, N

t = specimen thickness, mm

d = specimen diameter, mm

The indirect tensile strength ratio (TSR), is calculated using the following formula:

$$\text{Tensile Strength Ratio (\%)} = \frac{S2}{S1} \times 100$$

Where,

S1 = Average Tensile Strength of the Dry Subset

S2 = Average Tensile Strength of the Conditioned Sub-Set

Table 7. ITS and TSR for Conventional mix and modified mix

Description	Sample No.	Specimen Thickness (t) mm	Specimen Diameter (D), mm	Ultimate Load (P), N	ITS, MPa $\left(\frac{2000P}{\pi tD}\right)$
		Conventional mix			
Indirect Tensile Strength at 25°C for 2 hrs (Dry)	1	63.87	102.04	8620	0.84
	2	64.56	101.44	8400	0.82
	3	64.81	101.88	9100	0.88
				Average(S1) =	**0.85**
Indirect Tensile Strength at -18°C for 16 hrs, 60°C for 24 hrs & 25°C for 2 hrs (Wet)	4	62.37	101.67	7320	0.74
	5	64.15	101.25	7270	0.71
	6	63.00	101.75	7760	0.77
				Average(S2) =	**0.74**

			Tensile Strength Ratio (%)= $\frac{S2}{S1} \times 100$ **=**		**87**
		GF modified mix			
Indirect Tensile Strength at 25°C for 2 hrs (Dry)	1	62.91	101.70	11830	1.18
	2	64.38	101.84	12270	1.19
	3	63.36	101.73	13170	1.30
				Average(S1) =	**1.22**
Indirect Tensile Strength at -18°C for 16 hrs, 60°C for 24 hrs & 25°C for 2 hrs (Wet)	4	62.37	101.67	11090	1.11
	5	64.15	101.25	11540	1.13
	6	63.00	101.75	11210	1.11
				Average(S2) =	**1.12**
			Tensile Strength Ratio (%)= $\frac{S2}{S1} \times 100$ **=**		**92**

4.4 Retained Stability Test

[20] Measurements are made of the mix's resistance to moisture using the Marshall samples. After the samples are submerged in a water bath at 60°C for 30 minutes and 24 hours, the stability is assessed.

$$\text{Retained Stability (\%)} = \frac{S2}{S1} \times 100$$

Table 8. Retained stability for Conventional mix and modified mix

Description	Sample No.	Stability (kN)
	Conventional mix	
Retained Stability Test at 60°C for 30 minutes	1	23.2
	2	22.8
	3	20.8
	Average (S1) =	**22.3**
Retained Stability Test at 60°C for 24 hours	4	18.9
	5	20.1
	6	21.9
	Average (S2) =	**20.3**
	Retained Stability (%) = $\frac{S2}{S1} \times 100$ **=**	**91.1**
	GF modified mix	
Retained Stability Test at 60°C for 30 minutes	1	25.1
	2	24.2
	3	23.9
	Average (S1) =	**73.2**
Retained Stability Test at 60°C for 24 hours	4	22.5
	5	24.2
	6	23.1
	Average (S2) =	**69.8**
	Retained Stability (%) = $\frac{S2}{S1} \times 100$ **=**	**95.36**

Tables 7 & 8 show that the TSR for the GF-modified mix is higher than the conventional mix by 5.7% and 4.6% respectively. As per [21], the glass fibers drawn from crushed waste glass can also be brought into the future recommendations of research in pavements. While it is found that the tensile characteristics of the bituminous mix have improved, the rutting test holds high importance in evaluating the pavement characteristics which can be extended as future research.

5. Conclusion

After performing various tests, it was found that the use of glass fibres in a bituminous mix gave appreciable results.

- Glass fibre of length 8 mm and about 0.3% by weight of aggregates was found to be optimum with optimum binder content of 5.79%.

- An increase in Marshall Stability values was observed with the addition of glass fibre to the bituminous mix compared to that of the Conventional Mix
- Retained Stability was more significant in the case of Glass Fibre added Bituminous Mix (95.36%) than in Conventional Mix (91.17%).
- The Tensile Strength Ratio obtained from the Indirect Tensile Strength Test for Glass Fibre added Mix (92%) was observed to be greater than that of conventional mix (87%).
- An increase of 5% of TSR and 4.66% of retained stability of Glass fibre modified bituminous mix proves to be a promising result which opens up avenues of further research in this area.

Acknowledgements

Due credits are offered to Highway Research Station, Guindy, Chennai for providing technical support, laboratory facilities and guidance given to carry out this research.

References

[1] M. S. Eisa et al, 2020, "Effect of adding glass fiber on the properties of asphalt mix", International Journal of Pavement Research and Technology., DOI: 10.1007/s42947-020-0072-6
[2] Shaleha I. Vahora1, Prof. C. B. Mishra., "Study of Retained Stability on Warm Bituminous Mix", ICRISET2017 (Kalpa Publications in Civil Engineering, vol. 1), pp. 144–149.
[3] Gupta, L., Suresh, G. (2018). Determination of Indirect Tensile Strength of Bituminous Concrete Mix Prepared Using Stone Dust and Cement as Filler Materials. In: Struble, L., Tebaldi, G. (eds) Materials for Sustainable Infrastructure. GeoMEast 2017. Sustainable Civil Infrastructures. Springer, Cham. https://doi.org/10.1007/978-3-319-61633-9_16
[4] Shahbaz khan, M.N. Nagabhushana, Devesh Tiwari, P.K. Jain, "Rutting in Flexible Pavement: An Approach of Evaluation with Accelerated Pavement Testing Facility",Procedia - Social and Behavioral Sciences, Volume 104, 2013, Pages 149-157, ISSN 1877-0428, https://doi.org/10.1016/j.sbspro.2013.11.107.
[5] R. S. McDaniel, Fiber additives in asphalt mixtures. Project No. 20-05, Topic 45-15. Transportation Research Board, Washington DC, USA, 2015.
[6] C. J. Slebi-Acevedo, P. Lastra-González, P. Pascual-Muñoz, D. Castro-Fresno, Mechanical performance of fibers in hot mix asphalt: A review, Constr. Build. Mater. 200 (2019) 756- 769.
[7] Raj Pratul, Kumar Satish, Varsha, Singh Ajay Kumar, "Experimental study on fiber reinforced bituminous concrete", Materials Today: Proceedings, Volume 46, Part 20, 2021, Pages 11077-11083, ISSN 2214-7853, https://doi.org/10.1016/j.matpr.2021.02.227.
[8] O.S. Abiola, W.K. Kupolati, E.R. Sadiku, J.M. Ndambuki," Utilisation of natural fibre as modifier in bituminous mixes: A review", Construction and Building Materials, Volume 54, 2014, Pages 305-312, ISSN 0950-0618, https://doi.org/10.1016/j.conbuildmat.2013.12.037.
[9] Mashaan, N.; Karim, M.; Khodary, F.; Saboo, N.; Milad, A. Bituminous Pavement Reinforcement with Fiber: A Review. CivilEng 2021, 2, 599–612. https://doi.org/10.3390/ civileng2030033.
[10] Gao, Huan, Zhang, Lizhe, Zhang, Dongbo, Ji, Tao and Song, Jun. "Mechanical properties of fiber-reinforced asphalt concrete: Finite element simulation and experimental study" e-Polymers, vol. 21, no. 1, 2021, pp. 533-548. https://doi.org/10.1515/epoly-2021-0057.
[11] Mai Lan Nguyen, Juliette Blanc, Jean Pierre Kerzreho, Pierre Hornych. Review of glass fibre grid use for pavement reinforcement and APT experiments at IFSTTAR. Road Materials and Pavement Design, 2013, 14 (Supplement 1), pp.287-308. ff10.1080/14680629.2013.774763ff. ffhal-00848357
[12] A. Mahrez, M. Karim, H. Katman, Prospect of using glass fiber reinforced bituminous mixes, J. Eastern Asia Soci. Transp. Studies 5 (2003) 794-807.
[13] S. H. Khanghahi, A. Tortum, Determination of the optimum conditions for gilsonite and glass fiber in HMA under mixed mode I/III loading in fracture tests, J. Mater. Civ. Eng. 30 (7) (2018) 04018130.

[14] A. Najd, Z. Chao, G. Ying, Experiments of fracture behavior of glass fiber reinforced asphalt concrete, J. Chang'an University (Natural Science Edition), 25 (3) (2005) 28-32.

[15] Penki, R., Rout, S.K., Das, A.K. (2024). Number of Aggregate Sizes and Aggregate Gradation Area: A Correlational Study. In: Dhamaniya, A., Chand, S., Ghosh, I. (eds) Recent Advances in Traffic Engineering. RATE 2022. Lecture Notes in Civil Engineering, vol 377. Springer, Singapore. https://doi.org/10.1007/978-981-99-4464-4_30

[16] Rolands Izaks, Liga Gebauere, Romans Kornisovs, Viktors Haritonovs, "Laboratory Evaluation of the Properties of Dense Graded Asphalt Mixtures containing Waste Glass Fibre and Crumb Rubber", The Baltic Journal of Road and Bridge Engineering, Volume 17, Issue 2,2022

[17] Wei, J.; Mao, X.; Xu, W.;Xi, C.; Yan, S.; Sun, T.; Hu, X.;Wang, Y.; Chi, F. ExperimentalResearch on the Effect of Fiberglasson the Performance of EpoxyAsphalt Concrete. Sustainability 2022,14, 14724. https://doi.org/10.3390/su142214724.

[18] Goel, G., Sachdeva, S.N. & Pal, M. Modelling of Tensile Strength Ratio of Bituminous Concrete Mixes Using Support Vector Machines and M5 Model Tree. Int. J. Pavement Res. Technol. 15, 86–97 (2022). https://doi.org/10.1007/s42947-021-00013-5

[19] Rosa Veropalumbo, Nunzio Viscione & Francesca Russo. Indirect Tensile Strength Method For Defining A Proper Practice Of Asphalt Mixtures Design, WIT Transactions on Engineering Sciences, Vol 124, © 2019 WIT Press doi:10.2495/MC190251

[20] Wang, Xiushan & Zhou, Hengyu & Hu, Xingxing & Shen, Senjie & Dong, Bowen. (2021). Investigation of the Performance of Ceramic Fiber Modified Asphalt Mixture. Advances in Civil Engineering. 2021. 1-10. 10.1155/2021/8833468.

[21] G. Prabhakaran, Saddam Musa Patvegar, J. Prakash Arul Jose, D. Gouse Peera, Bhavesh Joshi, B. Ganeshkumar, "Study of the impact in bituminous mix using crushed waste glass", Materials Today: Proceedings, 2023, ISSN 2214-7853, https://doi.org/10.1016/j.matpr.2023.07.053.

Novel approach for strengthening of column by external wrapping with Carbon Fiber and Internally Embedded Damper System for resilient infrastructure

Murughendra Hiremath, A.K Roopa*, A. M. Hunashyal and Gourish Kallur

School of Civil and Environmental Engineering, KLE Technological University, Hubballi 580031, India

*E-mail: roopa.kuri@kletech.ac.in

Abstract. Columns as major structural components, lose strength due to sudden and accidental impacts. Such impacted columns can be restored by retrofitting, which increases their load-carrying capacity and lifespan, rather than reconstructing them. The common method of retrofitting involves externally wrapping the columns with fibre sheets. Numerous research studies have been conducted on enhancing the load-carrying capacity of a column by jacketing it with Fiber Reinforced Polymer (FRP). However, there is a lack of research on retrofitting columns internally to overcome vibrations produced due to impact loading. This paper addresses the retrofitting of a column both internally and externally. The external retrofitting is done by CFRP jacketing, and the internal retrofitting is achieved by developing a damper system to overcome damping forces. The experimental work aims to compare the load-carrying capacity of columns retrofitted internally with dampers and externally with CFRP jackets, with a plain concrete column. Dampers are evaluated based on energy absorption in comparison to plain cement mortar. It was found that energy absorption increased by 50% when the concrete was reinforced with 1mm chopped carbon fibers. Such reinforced concrete behaved as dampers, when placed inside the column to absorb the impact energy leading to a 9% increase in load-carrying capacity. The maximum load-carrying capacity was observed in columns wrapped with CFRP and damped with dampers, showing an increase of 25%. Finite Element Analysis is carried out to compare the experimental results and evaluate the damping property. Retrofitting a new column can enhance its initial load-carrying capacity, this helps in reducing the cross-section and percentage of steel usage in columns, contributing to sustainability. Retrofitting approaches to achieve 9th, 12th, and 13th sustainable goals, this leads to a significant contribution in the construction sector.

Keywords: Dampers, retrofitting of column, Carbon fibre, Fiber Reinforced Polymer(FRP), Sustainable Development Goals,

1. Introduction

Columns are the most important structural components of a structure, as they help transfer load from the superstructure to the substructure. The hierarchy of load transfer in a residential building is from the slab, to the beam, to the column, to the footing, and finally to the soil. When such structural members start to degrade and lose strength, they can be refurbished to maintain their original integrity, instead of demolition and reconstruction. This refurbishment process is known as retrofitting. In this method, materials are bonded externally in the form of tows, fabrics, plates, strips, and jackets. The most used retrofitting technique is jacketing. Steel jackets, reinforced concrete jackets, and fiber-reinforced polymer composite (FRP) jackets are the most popular types of jackets.[1],[2]

Since 1960, the construction industry has adopted Fiber-Reinforced Polymers (FRPs) in various technological aspects such as pressure pipes, tank lining, roof construction, and also in infrastructures like bridges, roads, and rail networks. The practice of using FRPs in the construction industry began in 1990. FRPs are further categorized based on the reinforced fiber matrix, such as carbon, glass, aramid, basalt, etc. If an FRP is reinforced with carbon, it is referred to as Carbon Fiber Reinforced Polymer (CFRP) [3-4]. CFRPs are known for their high mechanical strength and resistance to various chemical attacks and corrosion, they are also easy to install. These sheets are wrapped and adhered to structures using a super strong adhesive, such as epoxy resin. [5-6]

The performance of concrete can also be improved by adding nanofibers. Carbon nanofibers have the potential to enhance strength, effectively hinder crack propagation in cement composites, and act as nucleating agents [7]. Reinforcing concrete with nanofibers can produce tougher concretes by interrupting crack formation as soon as it is initiated. Carbon fibers, used in conventional fibre-reinforced concrete, have been found to be ideal reinforcing materials. They have established themselves as suitable materials for retrofitting and strengthening purposes. Hence, chopped carbon fibre can be reinforced in concrete to check load-carrying capacity and dynamic stiffness [8,9]

Chopped Carbon fiber with carbon fiber wrap has enhanced concrete strength by 44% as per [10,11] and it was noticed that liquid epoxy resin was better in adhering twill weave carbon fiber and concrete. Additional confinement is to be provided at the middle one-third of the height of the column to improve the ductility and strength because a buckling effect was observed in columns [12-15]. A Comparison was made between the ultimate load of controlled and wrapped columns with varying eccentricity in, it was found that load carrying capacity increased by 22.5% to 37.2% with increasing eccentricity, while the maximum deflection increased by 24% to 15%, However, the ultimate load decreases with increasing load eccentricity, and the maximum deflection increases with increasing load eccentricity [16,17]. The column loses strength due to corrosion, CFRP wrap was used to strengthen concrete columns they found that increasing CFRP wrap can eliminate the effect of pre-damage on the ultimate strain and increase load-carrying capacity [18-19]. Much research primarily focuses on external retrofitting, with little emphasis on the concept of internal damping. Dampers in columns enhances the resilience of buildings and infrastructure to seismic events by promoting the development of safer and more resilient infrastructure. This is especially crucial in areas that are prone to earthquakes, as employing dampers may significantly reduce the chance of structural failure and damage, protecting vital infrastructure [20,25].

Retrofitting a column also fulfils the sustainable goals for sustainable development. Sustainable development refers to meeting present needs without harming the future and surrounding nature [26,27]. To achieve such development, many conferences and goals have been set. These goals aim to balance economic development, environmental protection, and social wellbeing. Recently, in September 2015, the United Nations (UN) General Assembly adopted 17 Sustainable Development Goals (SDGs) as an integral part of the 2030 Agenda for Sustainable Development. These 17 goals were designed to build upon and broaden the scope of the earlier Millennium Development Goals (MDGs), which expired at the end of that year. [29- 30]. Each of the 17 Sustainable Development Goals (SDG) has a unique purpose. Moreover, the use of cement carbon fibre composite dampers in columns for retrofitting applications aligns with SDG 9 (Industry, Innovation, and Infrastructure) and SDG 11 (Sustainable Cities and Communities) by enhancing the safety and sustainability of urban areas. Sustainable building construction involves designing, building, and maintaining structures in a such way that, it remains ecological, and resource-efficient [31,32]. It attempts to provide conditions that are pleasant and healthful for inhabitants while minimizing the environmental effect of buildings throughout the lifespan.

The challenge is to meet them by the year 2030, This research helps to meet the 9th, 12th, and 13th SDGs, which serve the purposes of poverty, sustainable consumption, and climatic change mitigation.

The literature does not sufficiently address the implementation of dampers as retrofitting in structural members. While most papers present experimental results, only a few utilize Finite Element Method (FEM) analysis to substantiate their results. The combined effects of carbon fiber wrapping and damping are not thoroughly discussed. The literature also falls short in discussing the benefits of retrofitting for sustainability basically the approach to achieve sustainable development goals through retrofitting techniques remains unexplored. Henceforth, through literature review objectives are set, to design and develop a cement-carbon fiber matrix damper, and evaluating its mechanical and microstructural efficiency. These dampers are embedded in columns to test under external loading. To assess the deformation of the damper within the column, finite element modelling is carried out. Finally, the study is to be approached towards meeting the Sustainable Development Goals set by the United Nations in the 2015 summit.

2. Experimental Setup

Columns of size 150mm x 150mm x 300mm were cast for experimentation, wooden frames were used as Molds. The raw materials of concrete were tested for specific gravity and water absorption, and designed as per IS recommendations.

2.1 Materials Used

The materials used for the experimentation include Pozzolana Portland Cement, Coarse Aggregate of size 10-12mm for columns and 5mm for dampers, Fine Aggregates (River Sand), bi-directional carbon fiber sheets, and chopped carbon fiber up to 1mm in length. An M20-grade concrete was designed as per IS 10262-2019, resulted in a mix proportion of 1:1.3:3 and a water-to-cement ratio of 0.45 to cast rectangular column and Cubes were Cubes cured for 28 days resulted in a compression strength of 24.6 MPa. Bi-directional carbon fiber sheets were procured for retrofitting. An epoxy-based resin and hardener were used as adhesive agents between the concrete and fiber.

i) Material Properties: Pozzolana Portland Cement, Coarse Aggregate, and Fine aggregates were tested for specific gravity and water absorption, Specific gravity test using a Specific Gravity bottle and kerosene for cement, Pycnometer for fine aggregates, and Buoyance Balance equipment for Coarse Aggregates. Water absorption using temperature-controlled Owen. Test results are shown in Table 1,

ii)

Table 1. Properties of Materials

Materials	Specific Gravity	Water Absorption
Cement	3.2	-
Fine Aggregates	2.30	2.15%
Coarse Aggregates	2.80	0.44%

ii) CFRP Sheets: Commercially available CFRP sheets were used in this experiment to wrap around the column that increases load carrying capacity, mechanical properties of sheets are shown in Table 2.

Table 2. Mechanical Properties of CFRP Sheets

Description	Properties
Weave	Twill Weave
Design Thickness	0.1mm
Areal weight	200gsm
Tensile modulus	230 Gpa
Package	1m^2/roll

2.2 Preparation of Sample

The different specification of rectangular columns was casted with dimension of 150 x 150 x 300 mm for compression test as per Table 3.

Table 3. Sample specification

Short form	Composition
C1	Plain Concrete Column/Controlled Column
C2	Concrete column with internally Embedded Damper
C3	Concrete column with wrapped CFRP sheets
C4	Damper Embedded and CFRP wrapped Column

M20 grade concrete was used to cast columns and M25 grade for dampers. The concrete was designed according to IS 10262-2019. Figure 1, describes the process involved in casting of column samples. C1 is a plain concrete column in which M20 grade concrete was used. C2 has dampers embedded in the column. Before casting the columns, dampers were cast using M25 grade concrete reinforced with chopped carbon fiber, each 1mm in length, added in a quantity equal to 3% of the cement weight. The dimension of the damper is 150 x 150 x 10mm, it completely covers the cross-section of the column. These dampers were cured for 7 days, then placed in the core of the column. To achieve a good bond between the damper and fresh concrete, styrene butadiene latex by Dr. Fixit was used with cement in a 1:1 ratio. After casting, these columns were cured for 28 days.C3 columns are wrapped with CFRP sheets, they were casted with M20 grade concrete and cured for 28 days. After curing and surface drying for one day, CFRP sheets were adhered using epoxy resin and hardener in a 9:1 ratio. C4 columns have both a damper and CFRP sheets, i.e., columns embedded with a damper and wrapped with carbon fiber.

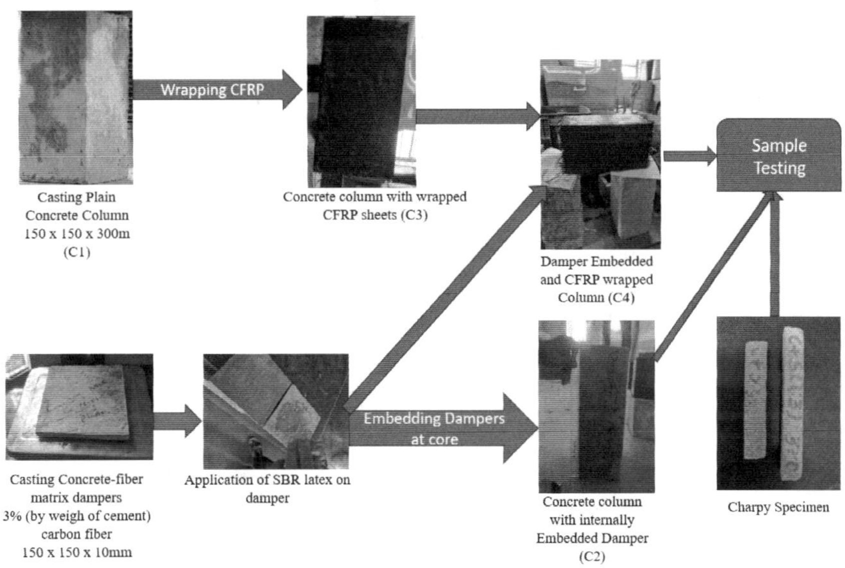

Figure 1. Sample Preparation

IOP Conf. Series: Materials Science and Engineering **1316** (2024) 012003 doi:10.1088/1757-899X/1316/1/012003

2.3 Experimental Testing

All columns, after casting and curing as per the procedure, were tested under a Universal Testing Machine (UTM) to compare the maximum load-carrying capacity of the columns as shown in Figure 2. A Charpy specimen with dimensions of 10 x 10 x 75mm was cast with reinforced cement mortar to check energy absorption. Universal Testing Machine (UTM) of 1000kN was used with a constant rate of loading, to observe the failure pattern and to know the maximum load carrying capacity of columns. For energy absorption capability Charpy specimen was tested in impact testing machine.

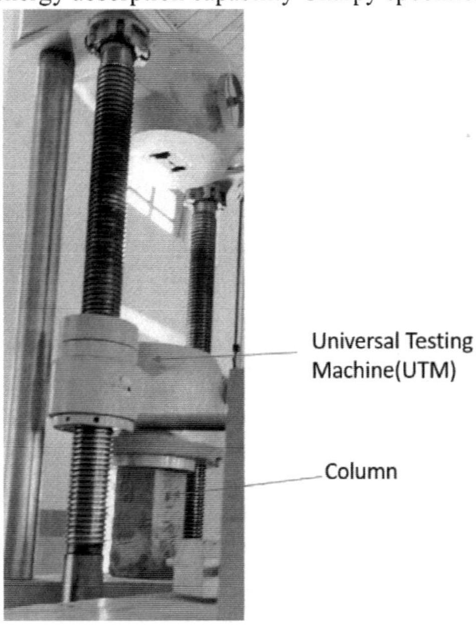

a. Universal Testing Machine

b. Impact Testing Machine

Figure 2. Testing Machine

3. Result and Discussion

The study results, load carried by columns of different compositions, the energy absorbed by the Charpy specimen, and the increased toughness of the column due to dampers. And correlate this with sustainability. The graph below illustrates the average values of load and toughness for each column type, providing a clear comparison of their performance.

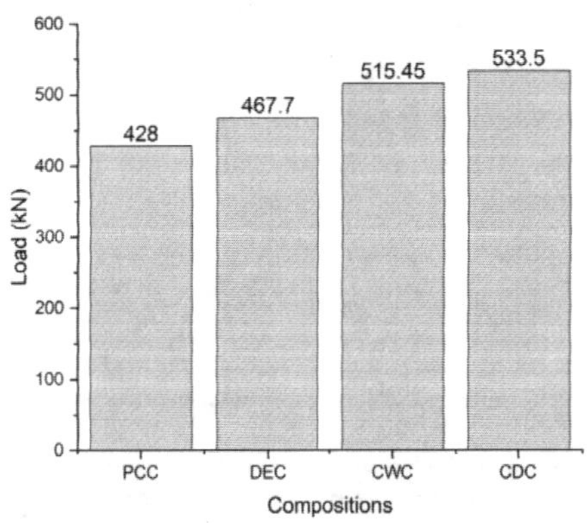

Figure 3. Load-carrying Capacity of different specification Column

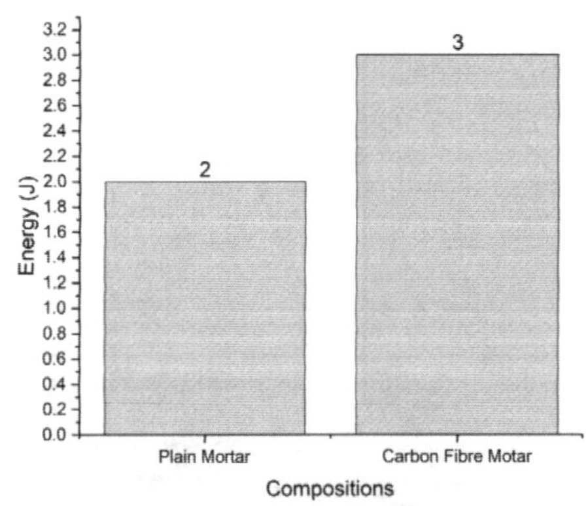

Figure 4. Impact Testing on Cement Mortar

IOP Conf. Series: Materials Science and Engineering **1316** (2024) 012003
doi:10.1088/1757-899X/1316/1/012003

Specimens tested under compression are presented in Figure 2. In testing, the specimens with carbon fiber wrap broke completely. Figure 3 shows the maximum load carried by the columns, compared with the C1 specimen percentage of increase is 9% for C2, 20.4% for C3, and 25% for C4. Here, the stiffness of the jacket in the transverse direction increases the resistance to lateral deformation of the column. As the stiffness of the jacket increases, the original column section experiences high confining pressure and hence experiences less lateral displacement, and carries load until the failure of the jacket. Carbon fiber jackets increased load carrying capacity whereas Carbon Fiber added damper enhanced the stiffness and carried the load in the form of Energy. Figure 4 represents the energy absorption of dampers through an impact testing machine, it represents the energy absorbed by the specimen before fracture. We can see a 50% increase in absorption because, Carbon fiber has a very high modulus of elasticity, when it is chopped and added to cement mortar, it eventually increases the elasticity of concrete/cement mortar. Therefore, carbon fiber-reinforced mortar has a high modulus of elasticity, which leads to increased energy absorption. This shows a huge potential in the energy absorption capacity of the fiber-added damper.

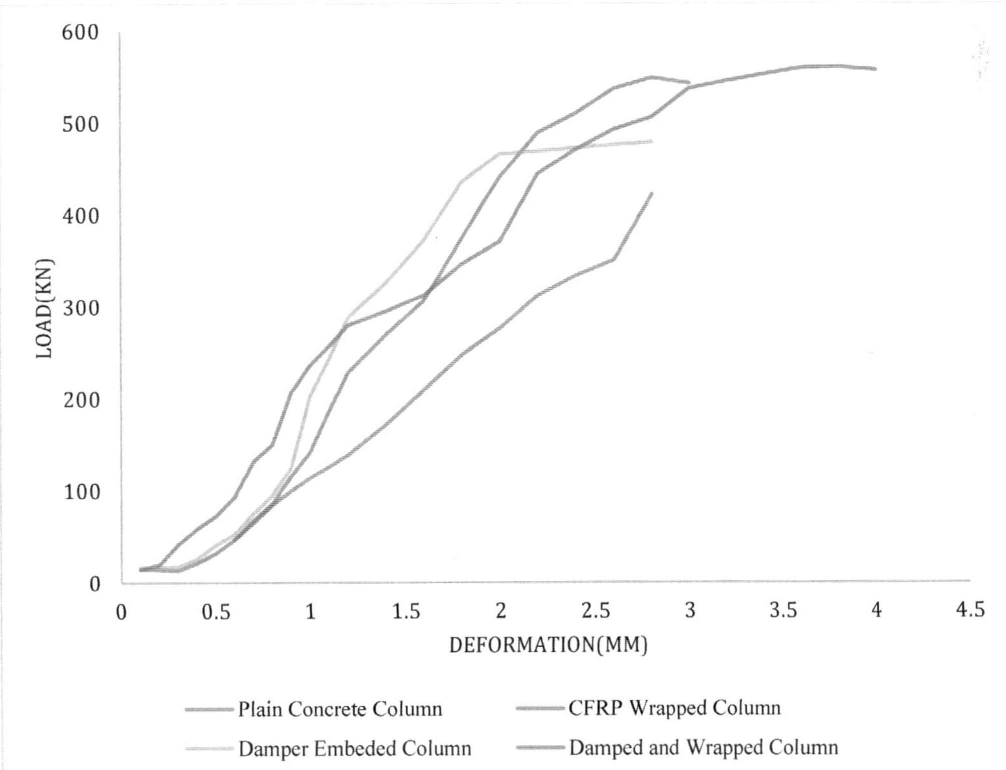

Figure 5. Load v/s Deformation of column

Figure 5 represents the load v/s deformation of each column, For the Plain Concrete column as a reference, we can see an increase in deformation for higher load. In CFRP Wrapped Column CFRP helps in increasing elastic region leads to carry load. The damped and wrapped Column has the highest deformation due to the damper and Compressive strength of CFRP wrap.Carbon fiber sheets possess a high Young's modulus, as Young's modulus increases Yield strength i.e elastic region of the specimen also increases, due to which the material can withstand the higher stress within the elastic region. Stress is directly proportional to load, hence by carbon fiber jacketing load carrying capacity of column will be increased.

In Wrapped columns, eventually load will be confined by the CFRP jacket, as the jacket fails the concrete experiences greater stress and deforms, which leads to failure.

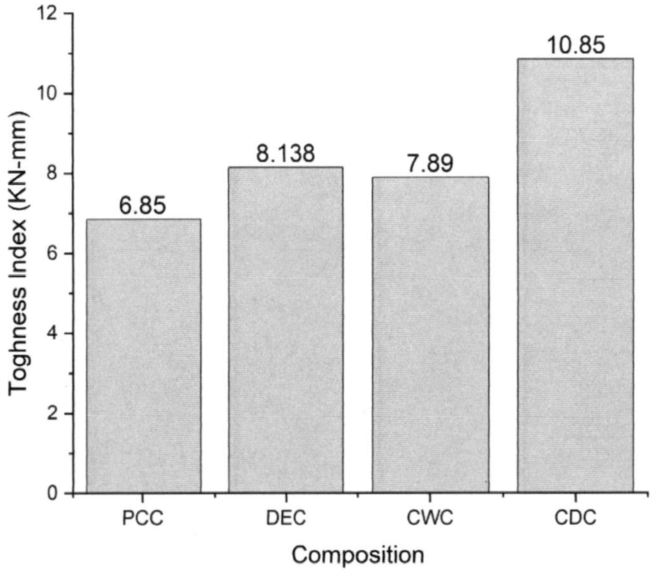

Figure 6. Toughness Index

Figure 6 shows the toughness index of various specimens, which is derived by calculating the area under the curve of a load-deformation curve. Carbon fiber has a higher stiffness due to its high modulus of elasticity. When it is embedded in concrete, it increases the stiffness of the concrete, The C1 Column is used as a reference, to compare other specimens. There is an increase in toughness as carbon content increases. The C4 column consists of the highest amount of carbon fiber in the form of chopped and wrapped, resulting in the highest toughness index.

4. Microstructural Analysis

Scanning Electron Microscopy (SEM) is conducted to understand the elementary composition of the specimen. Figure 7 presents the SEM analysis of cement carbon fibre damper. Due to the addition of carbon fiber into the cement forms a dense matrix. As Portland cement is a porous material, the addition of carbon fibre acts as bridge between the voids in cement mortar, hence increasing the strength of the matrix. This results in an increased load-carrying capacity and a decrease in shrinkage and cracks. Images A and B indicate that carbon fibers are well adhered to the CSH gel this indicates that carbon fiber helps to strengthen the concrete and Carbon fibers have a very high elastic modulus, By embedding them in concrete with C-S-H gel the modulus of elasticity of the concrete is also enhanced. This matrix results in increased damping properties compared to plain concrete dampers [33-34].

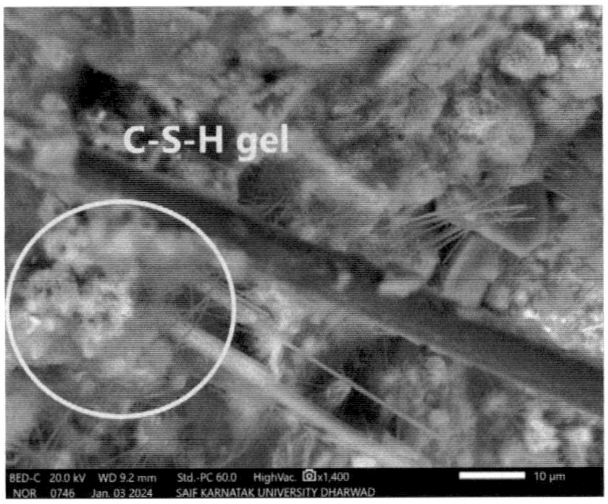

7a .C-S-H gel

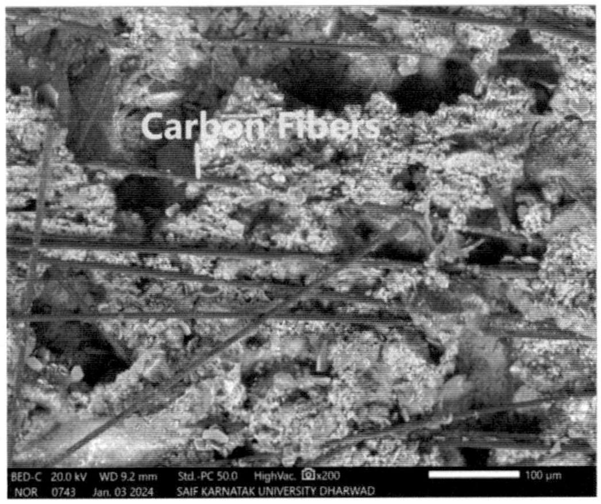

7b. Carbon fibers

Figure 7. SEM Analysis

5. Sustainable Development

Sustainable development of resilient structures with minimum materials involves designing and constructing buildings that are both environmentally friendly and capable of withstanding various hazards, such as earthquakes and extreme weather events. For sustainable development, the use of minimum materials involves optimizing the design and construction of buildings and infrastructure to reduce resource consumption and environmental impact. From the above study, compared to a plain concrete column load carrying capacity of a damped and wrapped column is higher hence by damping and wrapping a plain column we can reduce the section area of the column, to carry the same load by which cement content can be decreased.

Table 4. Sample specification

Sample	Composition	Load Carrying Capacity
C1	Plain Concrete Column/Controlled Column	468 kN
C4	Damper Embedded and CFRP wrapped Column	560 kN

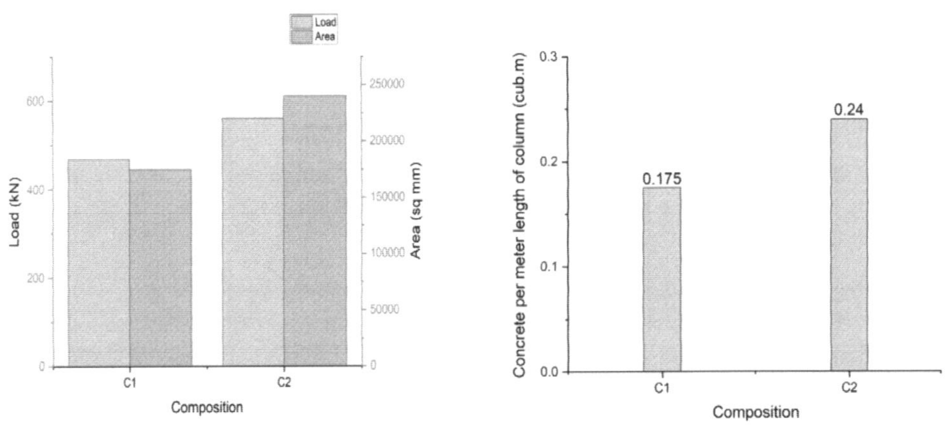

Figure 8 Reduction in Concrete and Cross section of column

From the Figure 8, it is observed that the controlled column carried a load of 468 KN, whereas by retrofitting the column was able to carry a load of 560 KN. If we design two columns as per IS 456-2000 such that one carries 468KN and the other carries 560KN, the cross-section of the column carrying 560KN is more compared to the column carrying 468 KN. The difference in area is about 40%, This implies by wrapping and damping a column we can increase its load-carrying capacity and also reduce cross-section area compared to normal design. The reduction of concrete is 0.0635m3 per meter length of column which concludes a reduction of 154.51 kg/m3 of cement in M20 grade concrete. This concludes Column design to carry 560 KN of load, can be achieved with the column dimension that carries 468 KN by retrofitting it internally and externally. By this 154.51 kg/m3 of cement can be reduced.

Table 5. SD goals as per UN

Sl. No.	Sustainable Development Goals	Purpose
SDG 9	Industry, Innovation, and Infrastructure.	Build sustainable infrastructure and foster innovation
SDG 12	Responsible Consumption and Production	Promote sustainable consumption and production.
SDG 13	Climate Action	Take urgent action to combat climate change and its impacts.

[a] source-https://sdgs.un.org/goals[33]

SDG 9: It aims to build resilient infrastructure, promote sustainable industrialization, and foster innovation. The study aligns to innovate a damper technique and introduce a new approach to building sustainable structures. These dampers reduce the grade of concrete by increasing deformation than design, which leads to a reduction in the section and material requirement of column casting. This, in turn, approaches to sustainable infrastructure development.

SDG 12: The goal encourages more sustainable consumption and production patterns through various measures. The Study contributes to the reduction of cement and steel, Given that the production of cement contributes approximately 8% of the world's anthropogenic CO2 emissions, reducing the usage of cement by minimizing section sizes will decrease the dependence on cement, aligning with this goal

SDG 13: This goal is to limit and adapt to climate change. As it is crucial to make changes in response to climate change, reducing carbon emissions from the construction industry (which accounts for around 40% of total emissions) can help reduce overall global carbon emissions, fulfilling this goal.

6. Finite Element Modelling

The finite element method is a technique used to analyse and understand physical phenomena and their effects on an object. It is predominantly used to study physical phenomena like wave propagation, structural or fluid behaviour, and thermal transport with the help of mathematics. In our study, it is used to know the deformation and behaviour of the damper under random vibration. ANSYS 22.2 version software is used for analysis.

6.1 Material Properties and Modelling

ANSYS software includes the properties of concrete and Carbon fiber, they are been modified accordingly. Modelling includes the dimensions of columns i.e. 150 x 150 x 300mm is set, dampers of 150 x 150 x 10mm, and Carbon fiber is also wrapped around the column with 0.5mm as thickness.

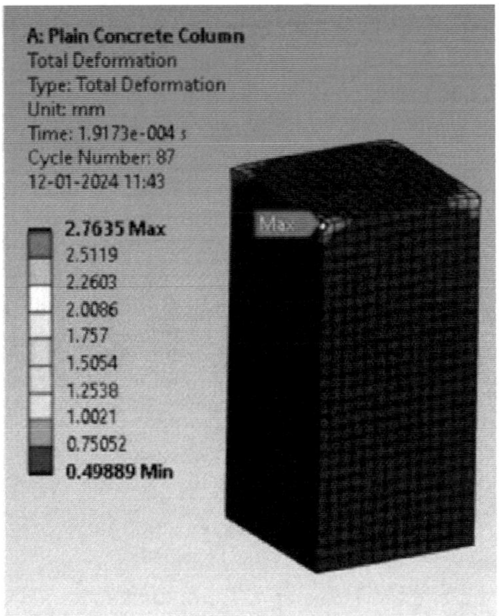

a. Plain Concrete Column

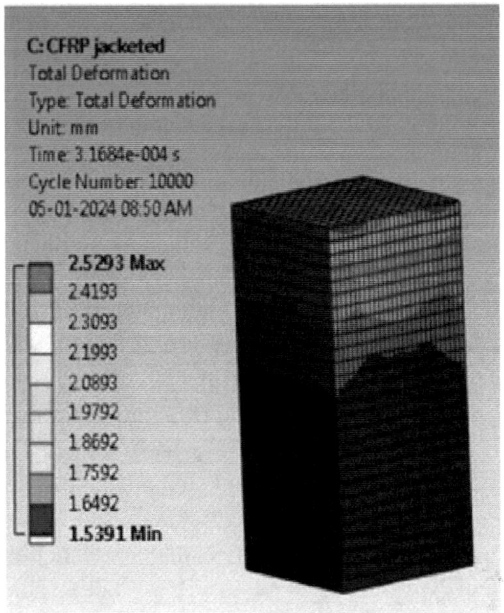

b. CFRP jacketed Column

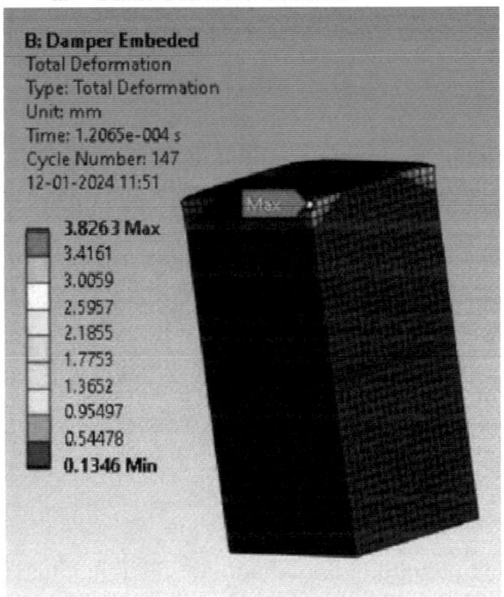

c. Damper Embedded Column

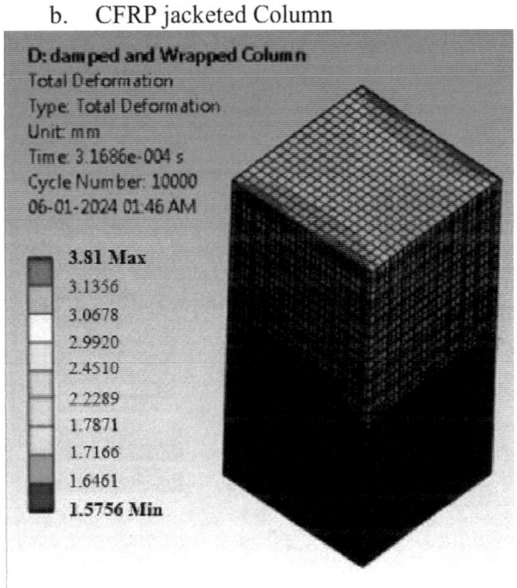

d. Damper Embedded and CFRP jacketed Column

Figure 9. ANSYS modelling od different beams

From the ANSYS results as per Figure 9, the maximum deformation found to be 2.7mm, 2.5mm, 3.6mm and 3.9mm respective. Edges being critical tend to deform and crack. The same failure pattern is seen in the experimental set up failed column. The maximum deformation is at the top most fibre of concrete. CFRP sheet helps in keeping concrete integrated and carries load. The maximum deformation is at the edges because eventually load tries to break the CFRP sheet but in model it is bonded with concrete as fixed object so there exist a large deformation, practically it tends to break sheet, it is because of point pressure on edges of column. The lower portion of the column is deformed less this indicates column can carry more load than assigned. In the figure 9c, the maximum deformation is at the top most fibre of concrete, Edges being critical tend to deform and crack. Eventually the failure point is same as plain concrete but the deformation is more due to embedded damper. While in Figure 9d, the CFRP sheet helps in keeping concrete integrated and carries load. The maximum is at the edges because eventually load tries to break the CFRP sheet but in model it is bonded with concrete as fixed object so there exist a large deformation, but practically it tends to break sheet.

6.2 Random Vibration Analysis

Due to the load, the column exhibiting vibrations is known as random vibration, the damper embedded in the column helps in overcoming this vibration. Analysis systems used are Random vibration analysis is Static structure to model and assign boundary conditions, Modal analysis to produce frequency from load, and finally Random vibration analysis In ANSYS such vibration can be tested on a column within its elastic limit, so the columns are modelled in a static structure. There exists a Modal Analysis system that creates six different deformations and calculates the frequency produced. The frequencies developed in modal analysis are analysed in a random vibration analysis system, through which the total deformation of the column is calculated. This deformation is caused due to frequencies, which are calculated using load. Therefore, columns are analysed for vibration due to applied load in the form of frequency. Through deformation and frequency power spectral density v/s Frequency graph is been obtained, here power is in terms of deformation hence the graph is between Deformation spectral density v/s Frequency.

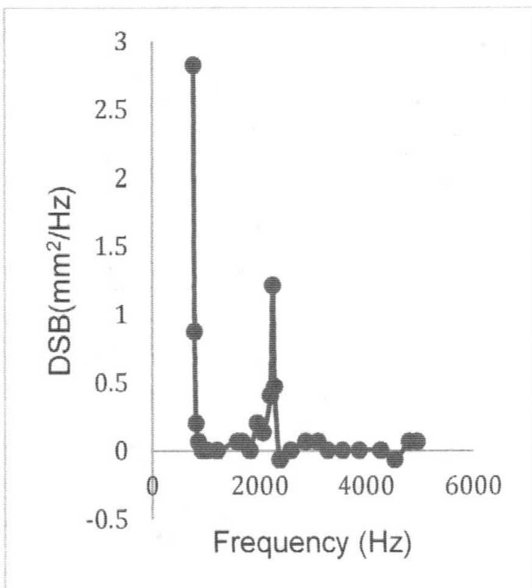

a. DSB v/s Frequency of C1

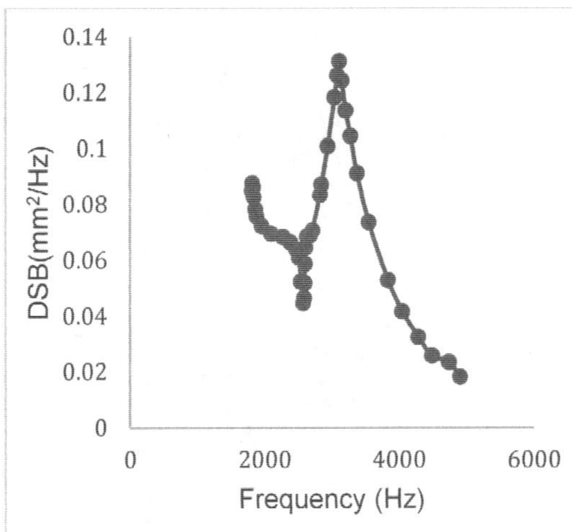

b. DSB v/s Frequency of C2

Figure 10. Random Vibration Analysis

The figure 10(a) represents a graph of a C1 column, the power to resist deformation is very high at the initial frequency. However, as the frequency increases, the resistance to deformation decreases. This implies that greater deformation is seen in the column, which can lead to failure, The Column fails quickly under random vibration showing a very large deformation. The figure10(b) represents the C2 column, the resistance to deformation initially decreases, which means smaller deformations will be observed in the column. As frequency increases deformation decreases, the increase in resistance to deformation is due to the damper carrying the load. The peak resistance to deformation occurs at 3100Hz, after which it starts to decrease. This implies that the column fails when the frequency produced exceeds 3100Hz, it is seen through declining of Damper Spectral Density in the graph.

7. Conclusion

The addition of 1mm chopped Carbon fiber to M25 grade concrete enhances the energy absorption of the concrete by 50%. When such concrete dampers are embedded into a column, the load-carrying capacity of the column increases by 9%. According to the literature review, Carbon Fiber Reinforced Polymer (CFRP) wrapping around columns enhances the load-carrying capacity. In our experiment, it increased by 20.4%. When a column is retrofitted both internally and externally, the load-carrying capacity increases by 25%. The random vibration of internally retrofitted columns shows that the columns were able withstand the vibration produced due to loading as compared to plain concrete column. The Scanning Electron Microscopy (SEM) study clearly concludes that the addition of carbon fiber fills up the pores, thereby decreasing shrinkage cracks, carbon fibers were also well dispersed that enhanced cement matrix strength. By retrofitting columns, there exist an approach to meet 9th, 12th, and 13th Sustainable Development Goals, i.e., building sustainable infrastructure by embedding dampers and wraps, creating sustainable cities by repairing old structures, and overcoming harsh climatic actions on structures by retrofitting. To enhance the performance of a damper, several strategies can be employed. One such strategy is to strengthen the bond between the damper and fresh concrete, this can be achieved by using epoxy-based latex. Another approach involves adjusting the length of the carbon fiber used in the casting of the damper. By experimenting with different fiber lengths, it is possible to find the optimum length that yields the maximum dampening force. Lastly, the percentage addition of chopped fibers can be altered and tested to determine the optimum percentage for fiber addition. By carefully adjusting these parameters, it is possible to improve the dampers performance significantly.

References

[1] M. Alhawamdeh and M. Alqam, "Behaviour Assessment of Reinforced Concrete Columns Externally Rehabilitated with Carbon Fiber-Reinforced Polymers (CFRPs) Subjected to Eccentric Loadings Behaviour Assessment of Reinforced Concrete Columns Externally Rehabilitated with Carbon Fiber-Rei," no. January, 2020.

[2] M. Nematzadeh, M. Mousavimehr, J. Shayanfar, and M. Omidalizadeh, "Eccentric compressive behavior of steel fiber-reinforced RC columns strengthened with CFRP wraps : Experimental investigation and analytical modeling," Eng. Struct., vol. 226, no. July 2020, p. 111389, 2021, doi: 10.1016/j.engstruct.2020.111389.

[3] P. Chotickai, P. Tongya, and S. Jantharaksa, "Performance of corroded rectangular RC columns strengthened with CFRP composite under eccentric loading," Constr. Build. Mater., no. xxxx, p. 121134, 2020, doi: 10.1016/j.conbuildmat.2020.121134.

[4] L. Antônio, M. Nunes, L. C. Aguiar, and E. Chahud, "Reinforcement of Concretes with Carbon Fiber Composite," no. March, 2014, doi: 10.5923/j.cmaterials.20140402.04.

[5] A. M. Hunashyal et al., "Experimental investigation of the effect of carbon nanotubes and carbon fibres on the behaviour of plain cement composite beams," vol. 3260, 2011, doi: 10.1080/19373260.2011.548160.

[6] F. Biermann, N. Kanie, and R. E. Kim, "Global governance by goal-setting: the novel approach of the UN Sustainable Development Goals," Curr. Opin. Environ. Sustain., vol. 26–27, pp. 26–31, 2017, doi: 10.1016/j.cosust.2017.01.010

[7] G. N. Narule and A. N. Bambole, "An experimental study on axial behavior of CFRP-strengthened RC rectangular columns with variable slenderness ratio," Asian J. Civ. Eng., vol. 22, no. 2, pp. 263–275, 2021, doi: 10.1007/s42107-020-00312-5.

[8] Vijayan, D. S. et al. Carbon Fibre-Reinforced Polymer (CFRP) Composites in Civil Engineering Application—A Comprehensive Review. Buildings 13, 1509 (2023).

[9] Narule, G. N. & Bambole, A. N. An experimental study on axial behavior of CFRP strengthened RC rectangular columns with variable slenderness ratio. (2020).

[10] Musale, Akshata, A. M. Hunashyal, and A. K. Roopa. "The Effect of Carbon Black, MWCNT and Carbon Fiber on Flexural behavior of cement nano composites." In E3S Web of Conferences, vol. 405, p. 03022. EDP Sciences, 2023.

[11] G. N. Narule and A. N. Bambole, “An experimental study on axial behavior of CFRP-strengthened RC rectangular columns with variable slenderness ratio,” Asian J. Civ. Eng., vol. 22, no. 2, pp. 263–275, 2021, doi: 10.1007/s42107-020-00312-5.

[12] Roopa, A. K., Dharwad, S., Hunashyal, A. M., & Ambale, D. (2024). Evaluating the Effectiveness of Titanium Dioxide Based Grouting Blended with Epoxy Coating for Retrofitting of Damaged Concrete Structures. In E3S Web of Conferences (Vol. 559, p. 04043). EDP Sciences.

[13] Lim, Jaehyeong, and Sang-Woo Kim. "Enhanced damping characteristics of carbon fiber reinforced polymer–based shear thickening fluid hybrid composite structures." Journal of Intelligent Material Systems and Structures 31, no. 20 (2020): 2291-2303.

[14] Nematzadeh, M., Mousavimehr, M., Shayanfar, J. & Omidalizadeh, M. Eccentric compressive behavior of steel fiber-reinforced RC columns strengthened with CFRP wraps: Experimental investigation and analytical modeling. Eng. Struct. 226, 111389 (2021).

[15] Losanno, Daniele, Ingrid E. Madera Sierra, Mariacristina Spizzuoco, Johannio Marulanda, and Peter Thomson. "Experimental performance of unbonded polyester and carbon fiber reinforced elastomeric isolators under bidirectional seismic excitation." Engineering Structures 209 (2020): 110003.

[16] Yagoub, Nouraldaim FA, Xiuxin Wang, Aamir Dean, Amr MA Moussa, and Elsadig Mahdi. "A novel resilient system of self-centering precast reinforced concrete walls with external dampers." Journal of Building Engineering (2024): 109107.

[17] Xu, T., Liu, J., Wang, X., Guo, Y. & Chen, Y. F. Behaviour of short CFRP-steel composite tubed reinforced normal and high strength concrete columns under eccentric compression. Eng. Struct. 205, 110096 (2020).

[18] Shin, D. & Kim, H. Cyclic response of rectangular RC columns retrofitted by hybrid FRP sheets. Structures 28, 697–712 (2020).

[19] Vijayan, Dhanasingh Sivalinga, Arvindan Sivasuriyan, Parthiban Devarajan, Anna Stefańska, Łukasz Wodzyński, and Eugeniusz Koda. "Carbon fibre-reinforced polymer (CFRP) composites in civil engineering application—a comprehensive review." Buildings 13, no. 6 (2023): 1509.

[20] S. R. Jalgar, A. M. Hunashyal, A. K. Roopa, M. A. Umarfarooq, S. N. Mathad, and M. S. Dhaduti, “Implementation of Cement-based nano composite Energy Absorption Damper to improve the damping properties of concrete and monitoring applications,” E3S Web Conf., vol. 455, 2023, doi: 10.1051/e3sconf/202345503019

[21] Wani, Zubair Rashid, Manzoor Tantray, and Javed Iqbal Sheikh. "Experimental and numerical studies on multiple response optimization-based control using iterative techniques for magnetorheological damper

[22] Selvaraj, Sivaganesh, and Mahendrakumar Madhavan. "Retrofitting of steel beams using low-modulus carbon fiber reinforced polymer laminates." Journal of Constructional Steel Research 167 (2020): 105825.

[23] Al-Zu'bi, Mohammad, Mizi Fan, and Lorna Anguilano. "Advances in bonding agents for retrofitting concrete structures with fibre reinforced polymer materials: A review." Construction and Building Materials 330 (2022): 127115.

[24] Roopa A K, Anand M Hunashyal. Evaluating Self-sensing Property of Carbon Fibre Cement Composite by Experimental Study and Finite Element Modelling For Structural Health Monitoring Applications doi:10.1088/1757-899X/1070/1/012041.

[25] Yu, P., Wang, Z., Lu, S. & Lai, P. Effects of Carbon Nanofibers and Carboxyl Functionalized Multi-Walled Carbon Nanotubes on Mechanical Damping Behavior of Cement Paste. J. Nanosci. Nanotechnol. 19, 163–169 (2018).
[26] F. Biermann, N. Kanie, and R. E. Kim, "Global governance by goal-setting: the novel approach of the UN Sustainable Development Goals," Curr. Opin. Environ. Sustain., vol. 26–27, pp. 26–31, 2017, doi: 10.1016/j.cosust.2017.01.010.
[27] Fei, Wenmei, Alex Opoku, Kofi Agyekum, James Anthony Oppon, Vian Ahmed, Charles Chen, and Ka Leung Lok. "The critical role of the construction industry in achieving the sustainable development goals (SDGs): Delivering projects for the common good." Sustainability 13, no. 16 (2021): 9112.
[28] Johnsson, Filip, Ida Karlsson, Johan Rootzén, Anders Ahlbäck, and Mathias Gustavsson. "The framing of a sustainable development goals assessment in decarbonizing the construction industry–Avoiding "Greenwashing"." Renewable and sustainable energy reviews 131 (2020): 110029.
[29] Mansell, Paul, Simon P. Philbin, and Efrosyni Konstantinou. "Delivering UN sustainable development goals' impact on infrastructure projects: An empirical study of senior executives in the UK construction sector." Sustainability 12, no. 19 (2020): 7998.
[30] de Almeida Barbosa Franco, Jacqueline, Ana Mariele Domingues, Nelson de Almeida Africano, Rafael Mattos Deus, and Rosane Aparecida Gomes Battistelle. "Sustainability in the civil construction sector supported by industry 4.0 technologies: challenges and opportunities." Infrastructures 7, no. 3 (2022): 43.
[31] Onat, Nuri Cihat, and Murat Kucukvar. "Carbon footprint of construction industry: A global review and supply chain analysis." Renewable and Sustainable Energy Reviews 124 (2020): 109783.
[32] Kynčlová, Petra, Shyam Upadhyaya, and Thomas Nice. "Composite index as a measure on achieving Sustainable Development Goal 9 (SDG-9) industry-related targets: The SDG-9 index." Applied Energy 265 (2020): 114755.
[33] Roopa, A.K.; Hunashyal, A.M.; Patil, A.Y.; Kamadollishettar, A.; Patil, B.; Soudagar, M.E.; Kalam, M.A. Study on Interfacial Interaction of Cement-Based Nanocomposite by Molecular Dynamic Analysis and an RVE Approach. Adv. Civ. Eng. 2023, 2023, 8404335
[34] Zhu, Hongbo, Haiyun Zhou, and Hongxiang Gou. "Evaluation of carbon fiber dispersion in cement-based materials using mechanical properties, conductivity, mass variation coefficient, and microstructure." Construction and Building Materials 266 (2021): 120891.

Novel Retrofitting Technique for Beam by External Wrapped with Glass Fiber and Internally Embedded Damper System for Sustainable Construction

Gourish Kallur[1], A.K Roopa[1*], A. M. Hunashyal[1] and Murughendra Hiremath[1]

[1] School of Civil and Environmental Engineering, KLE Technological University, Hubballi 580031, India

*E-mail: roopa.kuri@kletech.ac.in

Abstract. Reinforced concrete structures are often exposed to extreme loads, such as those from sudden and accidental impacts. This has led to an increasing interest in strengthening these structures, improving their fatigue performance, and extending their service life, particularly for components like beams. A more sustainable approach to maintaining their functionality involves strengthening and repairing damaged components. Therefore, glass fiber polymers are ideal reinforcements for retrofitting due to their high tensile strength and low cost compared to other polymer substitutes This paper presents the performance of plain concrete beams reinforced externally with glass fiber sheets (GFS) and dampers embedded internally. A unique methodology has been adopted to improve the adhesion between the fiber glass sheets (GFS) and the concrete surface. Two adhesive components utilized are epoxy resin (ER) and epoxy hardener (EH), mixed in a ratio of 9:1. Internally embedded dampers are devices used to resist lateral forces on structures, particularly during impact or sudden loading. The adopted damper technique involves incorporating chopped Glass Fiber during the casting process of the dampers. These dampers are subsequently embedded into beams at point of failure. After the curing of 28 days, all the cast beam compositions undergo flexural test, and dampers are tested for compressive strength. After the dampers have undergone compression testing, a microstructural analysis is conducted using SEM (Scanning Electron Microscope). For further details on stress formation in beams, finite element analysis of Ansys is used to model beams for all beam compositions. Also, sustainability goals are addressed by reducing cement usage in construction by introducing retrofitting in older buildings and using this system to improve the design of newer buildings by reducing overall section sizes

Keywords: Glass Fibre, Sustainable Development Goals, Retrofitting of Beam, Internal Damping,

1. Introduction

Reinforced concrete structures are often exposed to extreme loads, such as those from earthquakes and accidental impacts. This has led to an increasing interest in not only strengthening these structures but also improving their fatigue performance and extending their fatigue life, particularly for components like beams[1-2].

The need for this arises from inadequate original design limits, construction mistakes, lack of maintenance, or alterations in the building's usage. Up until now, several fortification methods have been employed, with reinforced concrete jacketing being the most common; however, the use of fiber-reinforced plastics (FRPs) as an alternative method has become increasingly popular. Despite this, there is no adequate code for the application of FRP strengthening techniques, and there is a lack of feedback on their use. Consequently, it is necessary to validate the specifications regarding the application of FRP strengthening techniques[3-5]. Due to its impressive tensile strength and modulus, glass fiber can effectively enhance the strength of organic composite materials, making it a popular choice in various industries. It has been extensively used not only in reinforcing organic composite materials but also in cement concrete.[5.-7]

Vibrations originating from structures or components are the main cause of issues in machine tools, aircraft, and automobile structures. These vibrations can be mitigated by incorporating damping into these structures. Vibrating systems dissipate energy in a variety of ways, and it is common for multiple energy dissipation mechanisms to occur at the same time. Damping, the process of energy dissipation, is typically categorized into two main types: material damping and structural damping[8]. Material damping refers to the energy dissipation that happens within the volume of the material. This type of dissipation is usually linked to internal changes in both macro and microstructures, ranging from molecular scales to effects on the crystal lattice, point defect relaxation, and thermo-elasticity[9]. On the other hand, structural damping happens due to the frictional force resulting from relative motion between components or intermittent contact among joints in the mechanical system. One method to mitigate vibrations is to use composites when building structures. Polymers, due to their higher viscoelastic property, are more effective in damping.[10-11]. Dampers are employed to counteract side-to-side forces on structures, especially during seismic events. Many researchers have developed precise analytical solutions for the free vibrations of tensioned beams equipped with a mid-position viscous damper. A different study has constructed a theoretical analysis model for a tight cable fitted with an internal rubber damper and an external viscous damper. The internal damper is presumed to be a high-damping rubber damper with flexible support, while the external damper is a linear viscous one.[13-16] A novel approach to augment the damping capacity of structures involves constructing them with composites, extending their lifespan.

The Sustainable Development Goals (SDGs), also referred to as the Global Goals, consist of 17 objectives established by the United Nations in 2015. The 2030 Agenda for Sustainable Development, which was embraced by all member states of the United Nations in the same year, offers a common framework for achieving peace and prosperity for humanity and the earth, both now and in the future. (Anon., 2015)[17-18]. Therefore, Sustainable Development Goals can be achieved through by installing dampers in the structures which are subjected to seismic loading to more resistant to earthquakes and other natural disasters, reducing risks and promoting safety, which is essential for achieving SDG 11. Additionally, installing dampers leads to achieving SDG 13 by making it possible to build lighter, more energy-efficient buildings, which could reduce their environmental effect and promote climate action[19-20].

This paper follows the basic principles of dampers and wrapping system. The integration of these systems with innovative techniques lends novelty to the study. The objective is to find new and effective ways to implement the retrofitting and making it novel. Research is conducted in different locations of damper placement and experimental investigations such as mechanical tests, microstructural analysis and finite element modelling are conducted. Also, reduction of

IOP Conf. Series: Materials Science and Engineering **1316** (2024) 012004 doi:10.1088/1757-899X/1316/1/012004

usage of cement is done to evaluate the efficiency of the retrofit and to align with sustainable development goals (SDG 9,12,13) set by United Nations by 2030.

2. Experimental investigation

2.1 Material used

The casting of beams was carried out using a concrete mix of 1:1.35:3 ratio and a water-to-cement ratio of 0.45, resulting in a 28-day cube strength of 24.6 MPa. The specifics of the beams are depicted in Figure 1. Bi-directional glass fiber-reinforced polymer fabrics, weighing 200 GSM, were employed for the reinforcement scheme. The laminates were prepared using a specialty two-part epoxy system, including resin and hardener components.

Table 1 represents the properties of materials used for experiment, such as Pozzolana Portland cement, coarse aggregate with a size range of 10-12 mm, and fine aggregate of river sand to cast the structural element. The Glass fibre is used for damping phenomena in structural member. The Glass fiber is used for damping phenomena in structural members. The glass fiber is chopped to a length of 1mm, and the glass fiber sheets are as per the mix design. Table 2 represents the properties of glass fiber.

Table 1. Properties of materials

Materials	Specific Gravity	Water absorption
Fine aggregate	2.30	2.15%
Coarse aggregate	2.80	0.44%
Cement	3.2	3.8%

Table 2. Properties of Glass fiber

Description	Properties
Density of fiber	200gm/m^2
Length of fiber	1 mm
Weave	Twill weave

2.2 Sample preparation

Four different compositions of samples are cast using above materials as per Table 2 are explained below. Figure 1 shows the process of preparation of samples.

Table 3. Sample specification

Sample specification	Composition
B	Plain concrete beam
BD	Plain concrete beam with 3% fiber added damper
BW	Plain concrete beam with externally wrapped GFRP sheets
BDW	Plain concrete beam with 3% fiber added damper and externally wrapped GFRP sheets.

2.2.1 Plain concrete beam

The materials, which have been tested for their physical properties, are employed in the casting a concrete beam to evaluate the compressive strength of the mix-designed concrete. The concrete mix, designed to meet M20 grade specifications, does not include the addition of glass fibers. A beam mould, measuring 150 x 150 x 300 mm, is utilized for casting the specimen. These beams are cast and subjected to a curing process for 28 days.

2.2.2 Concrete beam with internally retrofitted GFRP damper

The materials used for casting the plain concrete beam are replicated in this experiment. The concrete mix, designed to meet M20 grade specifications, is enhanced by adding glass fibers. These fibers, each 1mm in length, are added in a quantity equivalent to 3% of the cement's weight to a damper of a specimen measuring 150 x 150 x 10 mm. Each specimen is individually cured for 7 days. A beam mould, with 150 x 150 x 300 mm dimensions, is used for casting the concrete specimen, including dampers within the beams. For bond between the damper and fresh concrete, styrene butadiene latex (Pidicrete URP) is mixed with cement in a 1:1 ratio. After casting, these beams are left to cure for 28 days.

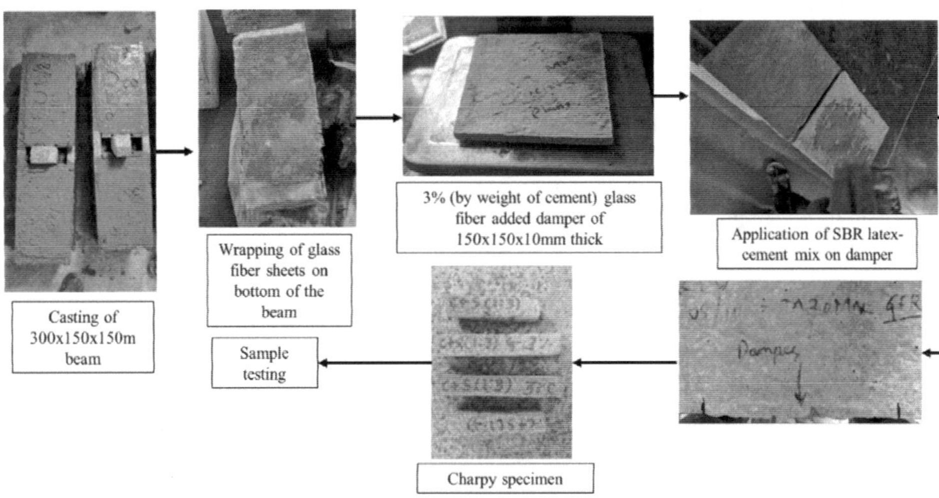

Figure 1. Preparation of samples

2.2.3 Concrete beam with externally retrofitted GFS

The concrete mix, designed to meet M20 grade specifications is used with the beam mould of dimension 150 x 150 x 300 mm is taken for casting the specimen. These beams are cast and cured for 28 days of curing period. After the curing period, use of GFRP sheets of the mentioned properties to cut a portion of 300 x 600 mm dimension and then wrap it around the beam using the epoxy binder.

2.2.4 Concrete beam with internally retrofitted damper and externally retrofitted GFS

The concrete mix, designed to meet M20 grade specifications, is enhanced with the addition of glass fibers. The fibers used are 1 mm in length and are added in a quantity equivalent to 3% of the weight of the cement to a damper of a specimen measuring 150 x 150 x 10 mm. Each specimen is individually cured for 7 days. A beam mould, with dimensions of 150 x 150 x 300 mm, is utilized for casting the concrete specimen. The same procedure of internally retrofitted beam composition is followed to achieve a bond between fresh and hardened concrete, which includes dampers within the beams. These beams are cast and left to cure for 28 days. After the curing period, GFRP sheets with specified properties are used for external wrapping. A 300 x 600 mm portion is cut and then wrapped around the beam using an epoxy binder.

2.3 Experimental set up

The three-point loading flexural test, a UTM with 1000kN capacity, is carried out on all samples to asses the flexural strength, maximum load and deflection [21-23]. For energy absorption capability of components impact test is conducted. Two samples of glass fiber of Izod and Charpy tests are casted of standard dimensions and tested. An impact test on the damper using the Charpy method was done. A specimen of size 10x10x75mm is tested for impact energy in an impact testing machine. The obtained energy value is the direct amount of energy required to rupture or break the specimen.

3. Result and discussion

Results are drawn from the experimental investigations of the specified beam and conclusion is drawn with the investigations.

3.1 Impact test on damper

The impact test assesses the energy absorbed by the Embedded glass fiber in the beam. As per Figure 2, the impact energy test conducted by the Charpy method shows the energy absorption of materials before fracture. This direct factor shows the dampening property of a damper with fibre addition. As per the graph, the difference in energy absorbed is 50% from the plain damper This shows a huge potential in the energy absorption capacity of the fibre-added damper.

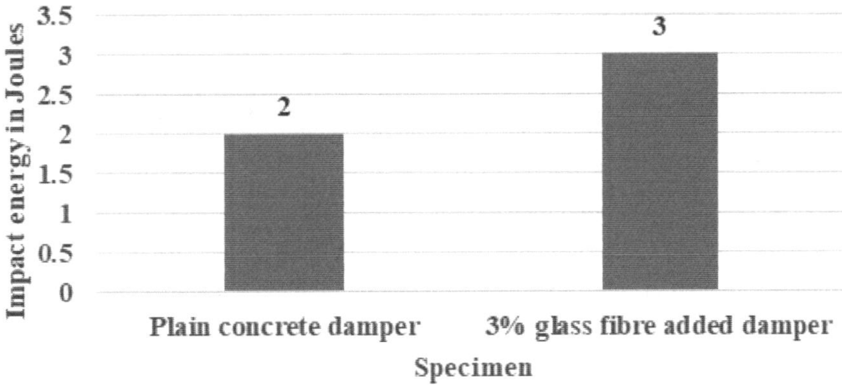

Figure 2. Impact energy of specimens

3.2 Flexural test

The three-point load bending testing is conducted using a universal testing machine with a 100-ton capacity to measure the load-carrying capacity of samples by plotting the load vs deformation graph. Figure 3 shows the normal concrete behaviour of a plain concrete damper as the stress strain increases linearly. After the yield point, it increases exponentially and fails at the ultimate point. For damper with wrapping, there is a linear graph that fails at the ultimate point. A beam with a damper shows an extended elastic zone compared to a plain concrete beam and resists loads with higher deformation. Lastly, the beam with damper and external wrapping has the highest deformation and load-carrying capacity as the damper holds the load by huge deformation, and later wrapping takes care of loading with reduced deformation after the dampers hit the yield point.

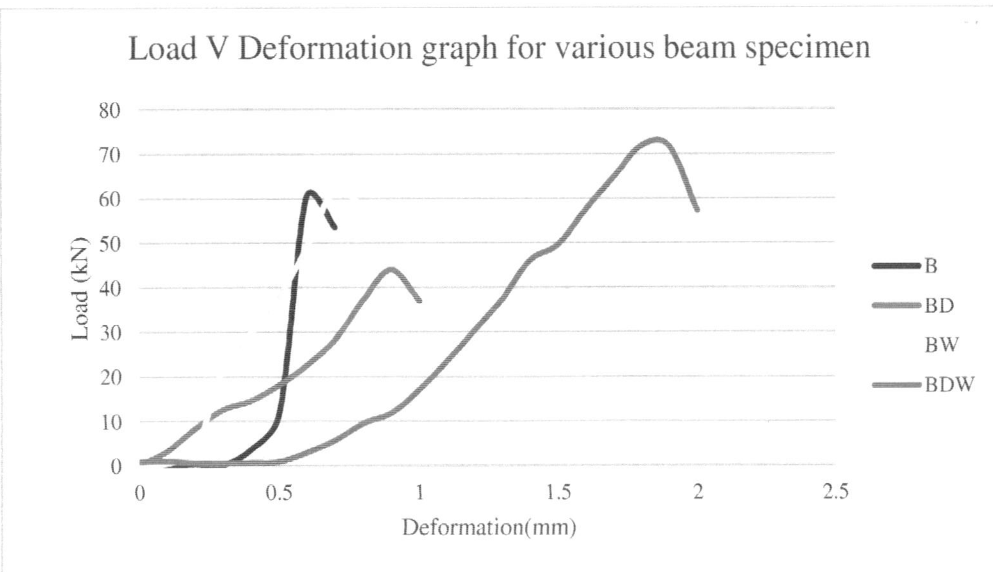

Figure 3. Load v/s deformation graph

IOP Conf. Series: Materials Science and Engineering **1316** (2024) 012004 doi:10.1088/1757-899X/1316/1/012004

The results derived from the graph indicate that the beam with the highest load-carrying capacity is one that is externally wrapped and internally embedded with a damper as shown in Figure 4. As the load increases, the damper strain also increases. However, GFS manages shear cracks and shear failure. The marginal increase in flexural strength in a beam embedded with a damper is attributed to shear failure in the beam. The crack propagation in the beam with the damper suggests that the crack propagates towards the near edge of the damper, leading to the failure of the beam.[22-24] The beam wrapped with GFS exhibits a significant increase in flexural strength. This is due to the high tensile strength of GFS, which is glued to the plane of the beam where the tensile stress is the highest. Shear stress in the beam can be managed by providing shear reinforcement, which is not possible in plain concrete specimens, or by modifying the specimen's dimensions. Figure 5 depcits the failure of samples under flexural loadig.

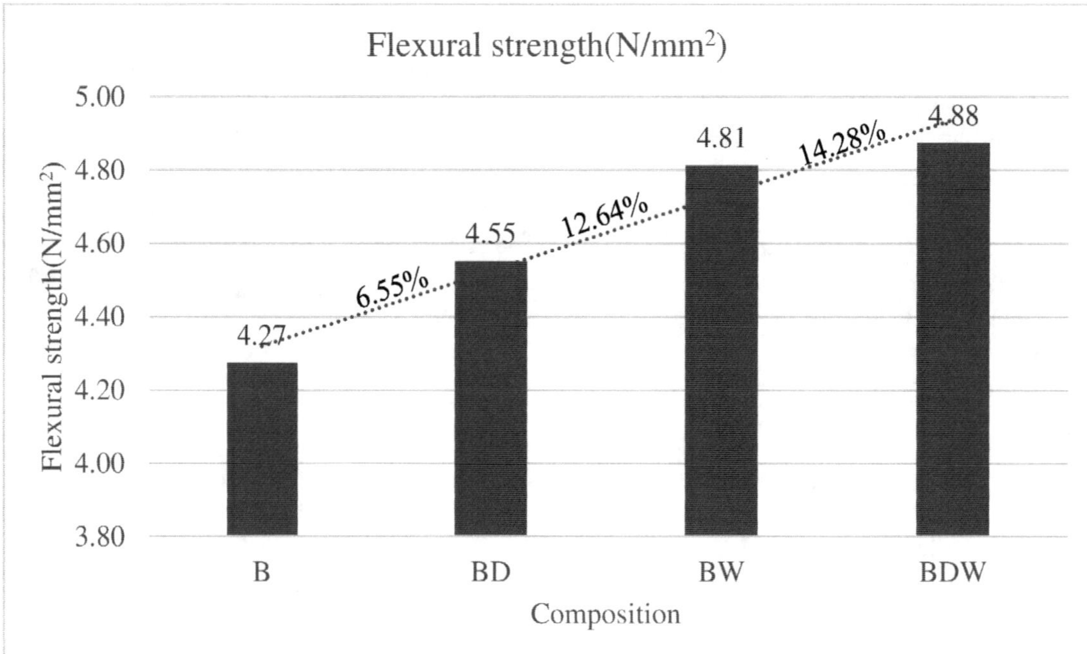

Figure 4. Flexural strength of various beams

Figure 5. Failure pattern of beam under flexural load

3.3 Toughness Index of damper

Figure 6 shows the toughness index of various specimens, which is derived by calculating the area under the load-deformation curve. As per the graph data, we can see an increasing trend. When it comes to the beam with damper, there is an increase in toughness index as the deflection increases as compared to the plain beam, and the same goes with the beam with damper and wrapping as the damper has a higher deformation and load-carrying capacity.

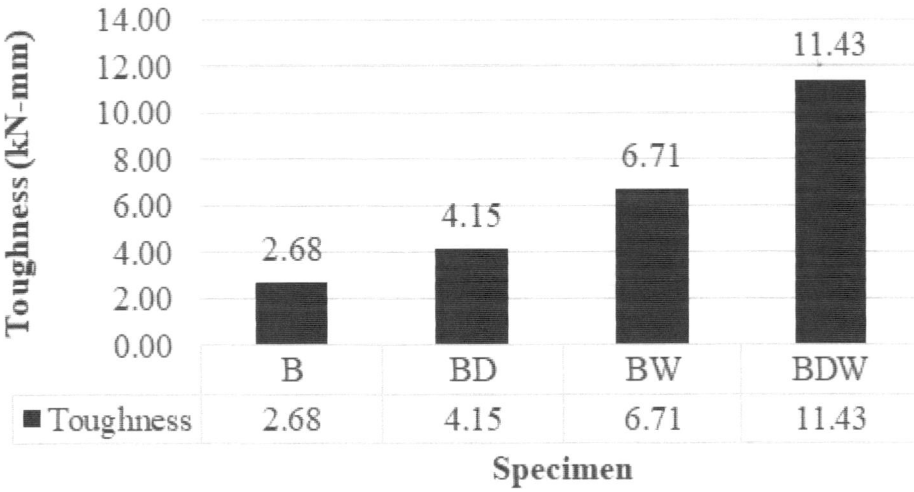

Figure 6. Toughness property of specimen

3.4 Microstructural analysis

SEM test results are carried out at Karnataka University Dharwad, India. SEM is conducted to understand the elementary composition of specimen. From the Figure 8 SEM image of damper embedded with glass fibers, it is evident that dense formation of the matrix is due to the addition of glass fiber into cement matrix as Portland cement is a porous material, addition of materials fibres fill up these pores hence the strength of matrix is increased, and shrinkage cracks are also

IOP Conf. Series: Materials Science and Engineering **1316** (2024) 012004 doi:10.1088/1757-899X/1316/1/012004

decreasing[25-26]. And the fibers behave as load dampers, hence increasing the ultimate load carrying capacity. In Figure 8 indicates the presence of glass fiber the embedding of fibers into the pores of concrete can be where C-S-H gel isn't formed. Increasing the density of the matrix. The black marker in both Figure 8 SEM image of damper embedded with glass fibers and Figure 9 indicates the formation of C-S-H. But the black spots in the Figure 9(b) indicates the pores or void due to improper formation of C-S-H gel. In SEM image represent the bonding of glass fibers and cement matrix and in Figure 7 represents the dispersion of glass fibers in the whole cement matrix

(7a)

(7b)

Figure 7a. SEM image of damper embedded with glass fibers and **Figure 7b.** SEM image of plain damper

3.5 Sustainable construction

Sustainable building construction is planning, creating, and maintaining structures in an approach that is conscious of the environment and resource It seeks to provide areas that are comfortable and healthful for residents while minimizing the environmental effect of buildings throughout the course of their lives. In the proposed work, through calculations as per IS 456-2000 limit state design it is found that through the experiment, we can reduce up to 30% of cement usage for per m^3 of concrete produced or i.e. 133.13 kg/m^3.

Table 4. Sample specification

Sample	Composition
B	Beam carrying load of 70kN
BDW	Beam carrying load of 88kN

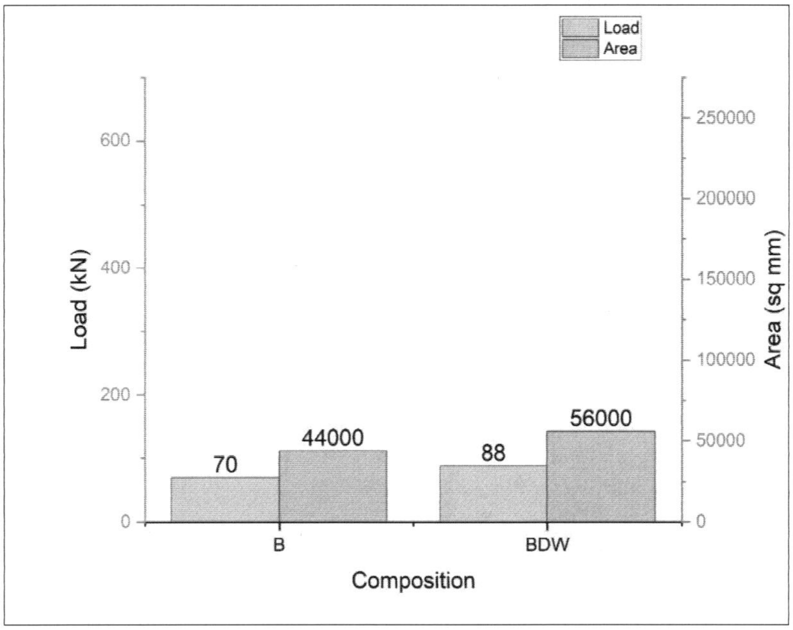

Figure 8 : Change of geometrical properties of sample

Table 5. SD goals as per UN

Sl. No.	Sustainable Development Goals	Purpose
SDG 9	Industry, Innovation, and Infrastructure.	Build sustainable infrastructure and foster innovation
SDG 12	Responsible Consumption and Production	Promote sustainable consumption and production.
SDG 13	Climate Action	Take urgent action to combat climate change and its impacts.

[a] source-https://sdgs.un.org/goals[33]

SDG 9 : It aims to build resilient infrastructure, promote sustainable industrialization, and foster innovation. The study aligns with the goal of innovating a damper technique and introducing a new approach to building sustainable structures. These dampers reduce the grade of concrete by increasing deformation than design, which leads to a reduction in the section and material requirement of beam casting. This, in turn, leads to sustainable infrastructure development

SDG 12: The goal encourages more sustainable consumption and production patterns through various measures. The Study contributes reduction of cement and steel, Given that the production of cement contributes approximately 8% of the world's anthropogenic CO_2 emissions, reducing the usage of cement by minimizing section sizes will decrease the dependence on cement, aligning with this goal

IOP Conf. Series: Materials Science and Engineering **1316** (2024) 012004 doi:10.1088/1757-899X/1316/1/012004

SDG 13 : This goal is to limit and adapt to climate change. As it is crucial to make changes in response to climate change, reducing carbon emissions from the construction industry (which accounts for around 40% of total emissions) can help reduce overall global carbon emissions, fulfilling this goal.

4. Finite Element modelling

The finite element method is a method employed to examine and comprehend the impact of physical phenomena on an object. It is primarily utilized to investigate phenomena such as wave propagation, the behaviour of structures or fluids, and thermal transport using mathematical principles. When addressing FEM problems on a computer, all problems are inputted as Partial Differential Equations (PDEs), which are typically hyperbolic, parabolic, or elliptic. Subsequently, boundary conditions or initial conditions are inputted into the software, and these inputs are assessed and solved[27].

4.1 Material properties and Modelling

For this study ANSYS 22.2 version software is used for finite element modelling. Materials data of concrete and GFS are provided to model the samples . Modelling is done by geometry tool and design modeller for all the specimens. The set dimension of 300x150x150mm is set. For all the specimen's size but damper of size 150x150x10mm is modelled and embedded in the beam. And for specimen with wrapping the glass fiber wrap around the beam is done with wrap tool.

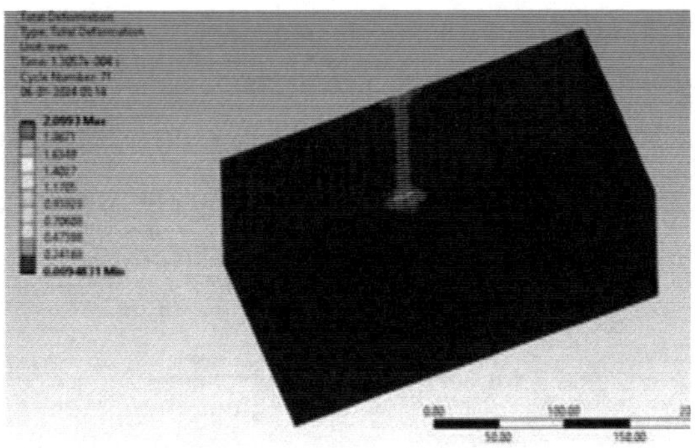

a. Plain concrete beam

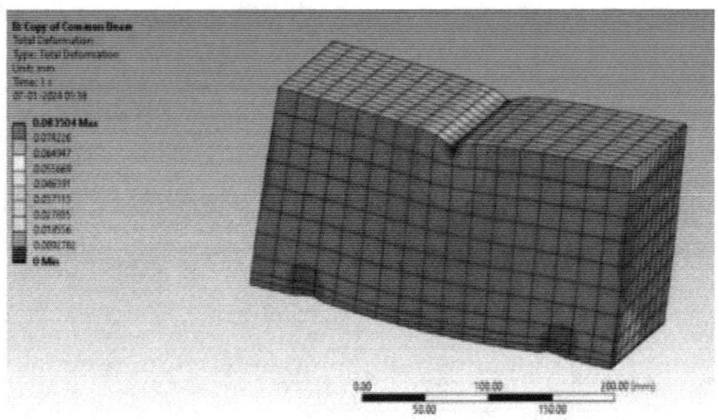

b. Plain concrete beam with damper

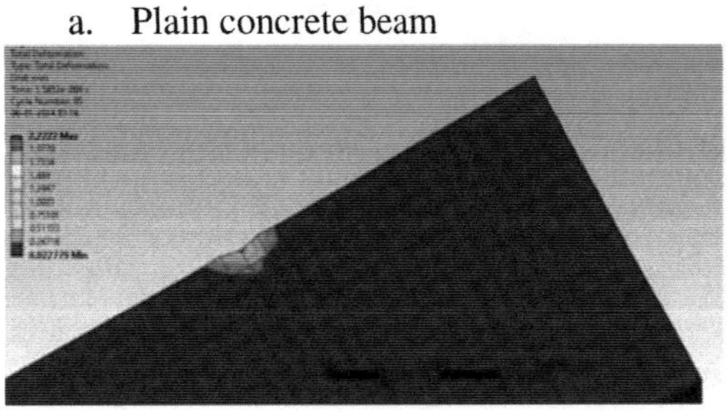

c. Plain concrete beam with externally wrapped GFRP sheets

d. Plain concrete beam with externally wrapped GFRP sheets and damper

Figure 9. ANSYS modelling od different beams

Figure 9 represents the results of the FEM analysis. The maximum deflection was 2.09mm, 1.95mm, 2.2mm, and 3.93 middle bottom portion of the concrete beam. The crack initiation takes place from the bottom of the beam . The same failure pattern is seen in the experimental set-up failed beam. From Figure 9b, the GF sheet helps keep concrete integrated and carries the load. The topmost portion of the beam is deformed less. This indicates that the beam can carry more load than assigned. In Figure 9C, the maximum deformation is at the bottommost fibre of concrete, the critical mid-portion, which tends to deform and crack. Eventually, the failure point is the same as plain concrete, but the deformation is more due to the embedded damper. GF sheet helps in keeping concrete integrated and carries the load.

4.2 Random Vibration

Due to the load, the beam exhibits vibrations known as random vibrations. The damper embedded in the beam helps to overcome these vibrations. In ANSYS, such vibrations can be tested on the beam within its elastic limit, so the beam is modelled in a static structure as shown in Figure 10. A Modal analysis system creates six different deformations and calculates the frequency produced. The frequencies developed in modal analysis are analysed in a random vibration analysis system, through which the total deformation of the beam is calculated. This deformation is caused due to frequencies, which are calculated using load. Therefore, the beam is analysed for vibration due to the applied load in frequency developed through six different ways.

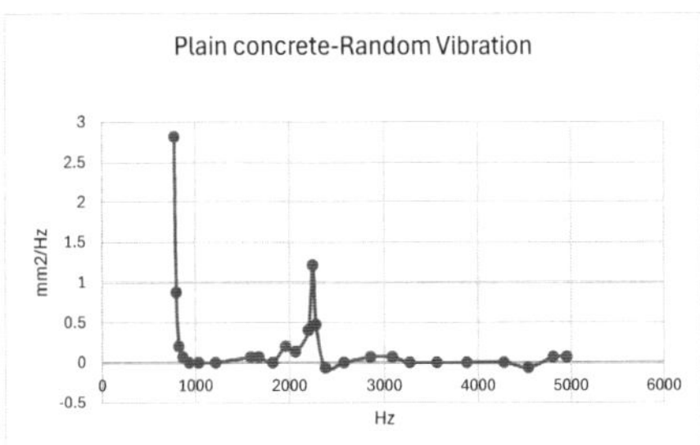

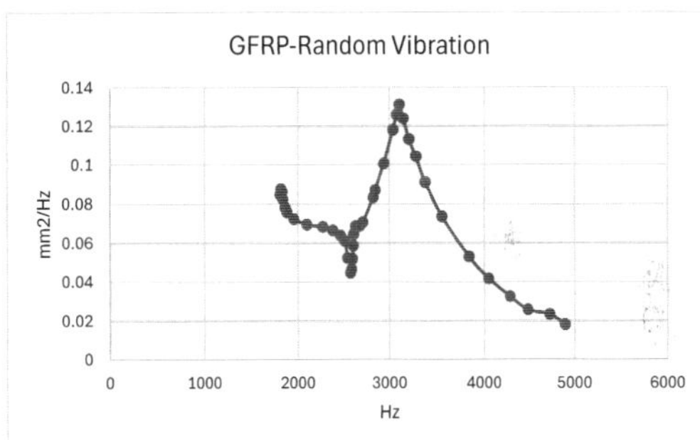

10a **10b**

Figure 10 a. Random vibration analysis of plain concrete beam and **10b.** Random vibration analysis of glass fibre reinforced beam

5. Conclusion and Future Scope

The application of GFS wraps has been demonstrated to be both efficient and effective in enhancing the load-bearing capacity and ductility of beams. These findings are not based on mere speculation, but on rigorous testing and analysis using ANSYS modelling, a widely accepted method in the field. When compared to control specimens, those retrofitted with GFS showed an increase in the ultimate load by 12.64%. The load for damped specimens increased by 6.55%, while the load for damped and wrapped beams increased by a 14.26%. In terms of ductility, the maximum deflection increased for beams that were damped, wrapped, and damped-wrapped with GFRP. The incorporation of 1mm chopped glass fiber into M25 grade concrete served as a damper to absorb load energy. Impact testing concluded that the energy absorption was by 3J. The glass fiber significantly enhanced the deformation capacity of the concrete. The deformation of beams under failure load in ANSYS modelling was a 2.09 to 3.39 millimetre. The difference in results was 10 to 20%. An SEM test was done to determine the elementary composition of the sample. The results clearly indicate that the addition of glass fiber filled up the pores, thereby increasing the strength of the matrix and reducing shrinkage cracks. The test also revealed that the glass fiber was well dispersed and bonded effectively with the cement matrix. The study fulfils the 9th, 11th,12th,and 13th Sustainable development goal. They are building sustainable infrastructure by embedding damper and wraps, creating sustainable cities by repairing old structures, and to overcome harsh climatic actions on structure by retrofitting.

The use of epoxy-based latex can enhance the bond strength between the damper and the rest of the beam. Various testing methods can be employed to ascertain the maximum load transfer. Adjustments can be made to the length of the glass fibre used in the casting of the damper to determine the optimal fiber length for the maximum damping force. Furthermore, the percentage of chopped fibers added can be varied and tested to find the optimal percentage for fiber addition. These modifications and tests can contribute to the overall performance and effectiveness of the damper in the construction process, equipping engineers and architects with the knowledge to optimize their designs.

References

[1] E. Ercan, B. Arisoy, and O. B. Ertem, "Experimental assessment of RC beam-column connections with internal and external strengthening techniques," Adv. Civ. Eng., vol. 2019, 2019, doi: 10.1155/2019/2828353.

[2] W. Bai, J. Zhang, P. Yan, and X. Wang, "Study on vibration alleviating properties of glass fiber reinforced polymer concrete through orthogonal tests," Mater. Des., vol. 30, no. 4, pp. 1417–1421, 2009, doi: 10.1016/j.matdes.2008.06.028.

[3] A. A. K. Sharba, H. D. Hussain, and M. Abdulhussain, "Retrofitting of RC beams using FRP techniques: a review," IOP Conf. Ser. Mater. Sci. Eng., vol. 1090, no. 1, p. 012054, 2021, doi: 10.1088/1757-899x/1090/1/012054.

[4] P. Kulkarni, A. Bhattacharjee, and B. K. Nanda, "Study of damping in composite beams," Mater. Today Proc., vol. 5, no. 2, pp. 7061–7067, 2018, doi: 10.1016/j.matpr.2017.11.370.

[5] L. Gemi, E. Madenci, Y. O. Özkılıç, Ş. Yazman, and A. Safonov, "Effect of Fiber Wrapping on Bending Behavior of Reinforced Concrete Filled Pultruded GFRP Composite Hybrid Beams," Polymers (Basel)., vol. 14, no. 18, pp. 1–20, 2022, doi: 10.3390/polym14183740.

[6] M. M. Attia, B. A. Abdelsalam, D. E. Tobbala, and B. O. Rageh, "Flexural behavior of strengthened concrete beams with multiple retrofitting systems," Case Stud. Constr. Mater., vol. 18, no. January, p. e01862, 2023, doi: 10.1016/j.cscm.2023.e01862.

[7] O. Rageh, M. A. El-Mandouh, A. H. Elmasry, and M. M. Attia, "Flexural Behavior of RC Beams Strengthened with GFRP Laminate and Retrofitting with Novelty of Adhesive Material," Buildings, vol. 12, no. 9, 2022, doi: 10.3390/buildings12091444.

[8] N. Eser, E. Töre, and İ. E. Bal, "Post-Earthquake Strengthening of RC Coupling Beams with GFRP Wrapping: Experimental Investigation," Materials (Basel)., vol. 16, no. 17, p. 6040, 2023, doi: 10.3390/ma16176040.

[9] Musale, Akshata, A. M. Hunashyal, and A. K. Roopa. "The Effect of Carbon Black, MWCNT and Carbon Fiber on Flexural behavior of cement nano composites." In E3S Web of Conferences, vol. 405, p. 03022. EDP Sciences, 2023.

[10] M. Valarezo Tandazo, J. Tu, and K. Gao, "Research on vibration control for glass fiber reinforced polymer reinforced concrete (GFRP RC) frames subject to seismic loading," Compos. Struct., vol. 261, no. September, p. 113290, 2021, doi: 10.1016/j.compstruct.2020.113290.

[11] Shafei, M. Kazemian, M. Dopko, and M. Najimi, "State-of-the-art review of capabilities and limitations of polymer and glass fibers used for fiber-reinforced concrete," Materials (Basel)., vol. 14, no. 2, pp. 1–45, 2021, doi: 10.3390/ma14020409.

[12] Shu, Z., You, R. and Zhou, Y., 2022. Viscoelastic materials for structural dampers: A review. Construction and Building Materials, 342, p.127955.

[13] S. R. Jalgar, A. M. Hunashyal, A. K. Roopa, M. A. Umarfarooq, S. N. Mathad, and M. S. Dhaduti, "Implementation of Cement-based nano composite Energy Absorption Damper to improve the damping properties of concrete and monitoring applications," E3S Web Conf., vol. 455, 2023, doi: 10.1051/e3sconf/202345503019.

[14] A P. Kulkarni, A. Bhattacharjee, and B. K. Nanda, "Study of damping in composite beams," Mater. Today Proc., vol. 5, no. 2, pp. 7061–7067, 2018, doi: 10.1016/j.matpr.2017.11.370.

[15] W. Bai, J. Zhang, P. Yan, and X. Wang, "Study on vibration alleviating properties of glass fiber reinforced polymer concrete through orthogonal tests," Mater. Des., vol. 30, no. 4, pp. 1417–1421, 2009, doi: 10.1016/j.matdes.2008.06.028.

[16] V. Dacol, E. Caetano, and J. R. Correia, "Comparative Study of Damping on Steel and Pultruded GFRP Beams.," Polymers (Basel)., no. In review-Submission at 2021-06-13, pp. 1–24, 2021.

[17] Fei, Wenmei, Alex Opoku, Kofi Agyekum, James Anthony Oppon, Vian Ahmed, Charles Chen, and Ka Leung Lok. "The critical role of the construction industry in achieving the sustainable development goals (SDGs): Delivering projects for the common good." Sustainability 13, no. 16 (2021): 9112.

[18] https://sdgs.un.org/goals

[19] Ogunmakinde, Olabode Emmanuel, Temitope Egbelakin, and William Sher. "Contributions of the circular economy to the UN sustainable development goals through sustainable construction." Resources, Conservation and Recycling 178 (2022): 106023.

[20] Gade, Anne Nørkjær, and Alex Opoku. "Challenges for implementing the sustainable development goals in the danish construction industry: building owners' perspective." In ARCOM 2020-Association of Researchers in Construction Management, 36th Annual Conference 2020-Proceedings, pp. 615-624. ARCOM, Association of Researchers in Construction Management, 2020.

[21] Roopa, A. K., A. M. Hunashyal, Fardeen A. Gudagi, and Mohammed Furquan Mulla. "Piezo-resistive Properties of Polymer based cement nano composites." In E3S Web of Conferences, vol. 455, p. 03013. EDP Sciences, 2023.

[22] S. R. R. Teja Prathipati and C. B. K. Rao, "A study on the uniaxial compressive behaviour of graded fiber reinforced concrete using glass fiber/steel fiber," Innov. Infrastruct. Solut., vol. 6, no. 2, pp. 1–14, 2021, doi: 10.1007/s41062-020-00438-0.

[23] D. Barkale and S. B. Thakare, "Parametric Studies on Structural Behaviour of Strengthened Beams Using Glass Fiber Reinforced Plastic," Adv. Sci. Technol. Res. J., vol. 16, no. 6, pp. 192–197, 2022, doi: 10.12913/22998624/156336.

[24] Ercan, B. Arisoy, and O. B. Ertem, "Experimental assessment of RC beam-column connections with internal and external strengthening techniques," Adv. Civ. Eng., vol. 2019, 2019, doi: 10.1155/2019/2828353.

[25] A. A. R. Amer, M. M. A. B. Abdullah, L. Y. Ming, and M. F. M. Tahir, "Performance and properties of glass fiber and its utilization in concrete - A review," AIP Conf. Proc., vol. 2030, no. November, 2018, doi: 10.1063/1.5066937.

[26] A.Raza, A. Rashedi, U. Rafique, N. Hossain, B. Akinyemi, and J. Naveen, "On the structural performance of recycled aggregate concrete columns with glass fiber-reinforced composite bars and hoops," Polymers (Basel)., vol. 13, no. 9, 2021, doi: 10.3390/polym13091508.

[27] Roopa, A.K.; Hunashyal, A.M.; Patil, A.Y.; Kamadollishettar, A.; Patil, B.; Soudagar, M.E.; Kalam, M.A. Study on Interfacial Interaction of Cement-Based Nanocomposite by Molecular Dynamic Analysis and an RVE Approach. Adv. Civ. Eng. 2023, 2023, 8404335

IOP Conf. Series: Materials Science and Engineering **1316** (2024) 012005 doi:10.1088/1757-899X/1316/1/012005

Accelerating Linear Congruential Generators with Carbon Nanotube Field-Effect Transistors

B Yogapriya, P Sahithi, Mathan N*, Muthiah M A and T Ravi

Sathyabama Institute of Science and Technology, Chennai, India

**Corresponding Author: mathan.ntmoorthy@gmail.com*

Abstract. In the era of technological advancements, safeguarding information privacy has become paramount across sectors like defense, finance, healthcare, and more. At the heart of these security measures lies cryptography, a field that necessitates the efficient processing and secure storage of data. Our project focuses on the optimization of Linear Congruential Generators (LCG), pivotal for cryptographic applications and pseudorandom number generation. Here it has been pioneered the implementation of LCG using Carbon Nanotube Field-Effect Transistor (CNTFET) technology, targeting the reduction of leakage power through the innovative application of sleep techniques. This advanced approach not only embodies a leap towards minimizing energy consumption but also enhances the overall performance and reliability of cryptographic systems. By integrating sleep modes into the CNTFET-based LCG design, our methodology significantly curtails leakage power without compromising the generator's efficiency or output quality. The successful implementation and synthesis of this design, employing cutting-edge CNTFET technology, underscore the potential for substantial improvements in power efficiency while maintaining high standards of security and randomness essential for cryptographic algorithms.

1. Introduction

In the realm of cryptographic systems, achieving optimal system performance while maintaining robust security is paramount. Hardware implementations of cryptographic algorithms are crucial for this, with modular arithmetic operations such as modular exponentiation, multiplication, and addition being central to the functionality of various encryption methods. The efficiency of these cryptographic algorithms significantly depends on the streamlined execution of these modular arithmetic operations. Our paper highlights the critical role of efficient hardware architecture in enhancing the performance of cryptographic operations, with a specific focus on the implementation of Linear Congruential Generators (LCG) for pseudo-random bit generation. LCGs, which are fundamental to algorithms like the modified dual-CLCG (MDCLCG), rely on basic arithmetic operations including binary addition, underscoring the need for an effective implementation strategy. Distinctly, our project emphasizes the MDCLCG as a secure and highly random pseudo-random bit generation method, capable of providing polynomial-time unpredictability for bit sizes beyond 32 bits. The architecture of MDCLCG faces challenges in scaling, particularly with increases in area and critical path delay as operand size expands[1]. To address these challenges, our study introduces an innovative approach by employing Carbon Nanotube Field-Effect Transistor (CNTFET) technology for the implementation of the LCG, alongside a comparative analysis with traditional MOSFET technology. Furthermore, we have incorporated a sleep technique aimed at reducing leakage power, thereby enhancing the energy efficiency of our cryptographic hardware design and to advance the performance and security of pseudorandom bit generation in cryptographic applications, offering insights into the benefits of modern semiconductor technologies in cryptography[2].

2. Linear Congruential Generator (LCG)

The most popular method for producing random numbers is a technique known as the linear congruential generator. Lehmer presented the concept in accordance with the subsequent equation in

$$X_{n+1} = (aX_n + c) \bmod m \qquad (1)$$

The modulus (m) ,multiplier(a) and increment (c) are crucial parameters in the configuration of a Linear Congruential Generator (LCG) that significantly affect its ability to produce a wide range of non-repeating values before reaching the modulus(m).The selection of these parameters must be done with precision to ensure the generation of a sequence of values that appears random and has a long period. According to the guidelines provided by Park and Miller , setting the increment (c) to 0 can yield any satisfactory results[3]. In this setup , the multipier(a) should be chosen from within the range of 2 to m-1 , and the modulus (m) should be a large prime number.Although the theoretical maximum cycle length of an LCG is the modulus (m), achieving this maximum length requires adherence to specific conditions.

1. The increment (c) should be relatively prime to the modulus (m), ensuring that every possible value in the range can eventually be generated.
2. The multiplier (a) minus 1 should be divisible by all primefactors of the modulus (m).
3. When the modulus (m) is a multiple of 4, the multiplier (a) minus 1 should also be a multiple of 4 This condition helps in maintaining the cycle length by aligning the generator's operation with the modulus's properties.

2.1. LCG circuit diagram

The circuit diagram of LCG operation overall is shown in Figure1 (seed is disregarded). This circuit requires multipliers, adders, comparators, and subtractors. The multiplier is used to multiply the preceding random number which is X, with the multiplier (a) and the output is given to an adder to add with c. The modulus m is then contrasted with it after that[4]. If the number is less than m or equal to m, the obtained result is considered as a random number. In either case, m is subtracted with the integer. The result produced after subtraction is then considered as a random number.

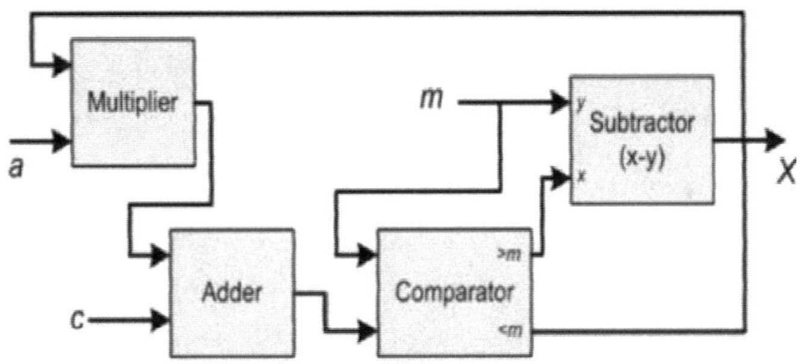

Figure 1. LCG circuit diagram

This circuit implements the Linear Congruential Generator (LCG) algorithm for generating pseudo-random numbers but eliminates the need for separate subtractor and comparator blocks. Instead, it relies on a multiplexer and clever use of the adder and buffer combination for efficient operation. The multiplexer selects between two possible inputs based on a control signal[5]. In this circuit, it chooses either the previous random number (stored in Buffer B4) or the initial seed value (Seed) depending on the "reset" signal. The multiplier unit multiplies the selected input (from the multiplexer) by the constant multiplier value (A) provided at the "A" port.The adder adds the result of the multiplication to the increment value (C) provided at the "C" port. However, it also serves a dual purpose.

Buffer B1 stores the current random number before it enters the adder.Buffer B2 stores the result of the multiplication in the previous cycle.Buffer B3 holds the sum of the multiplied value and the increment.Buffer B4 stores the final random number generated in the previous cycle.

2.2. LCG operation

When the "reset" signal is HIGH (and "enable" is LOW), the multiplexer selects the Seed value, which is then written into Buffer B4. This clears any previous values and sets the starting point for the random number generation.When the "enable" signal goes HIGH (and "reset" is LOW), the circuit starts generating random numbers[6].With each clock pulse the previous random number from Buffer B4 is fed into B1.With each clock pulse the previous random number from Buffer B4 is fed into Buffer B1 .The selected value is multiplied by A and stored in Buffer B2.The value in B2 is added to C and stored in Buffer B3.

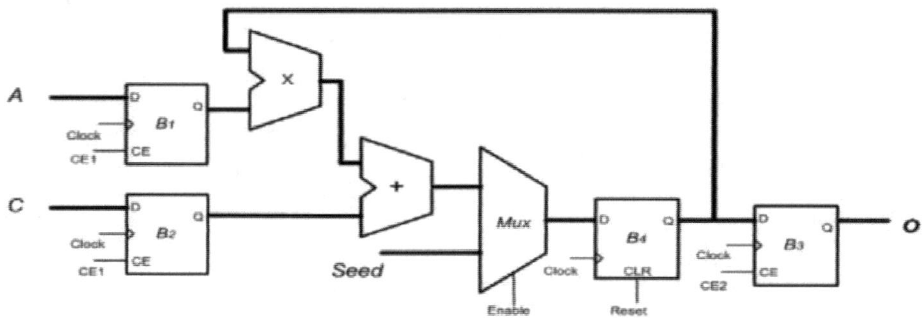

Figure 2. LCG block diagram

Instead of comparing B3 with the modulus (m), the adder wraps around if the sum exceeds m. This means it automatically subtracts m from the sum, effectively achieving the comparison and subtraction in one step.The final random number (now within the range) is written to Buffer B4, overwriting the previous one. This process repeats with each clock cycle, generating a new random number based on the previous one. The sequence depends on the initial seed, A, and C values.The LCG circuit is analyzed using the 130nm and 32 nm MOSFET and 32nm CNTFET technology[7][8] .

3. CNTFET technology

Carbon Nanotube Field-Effect Transistor (CNTFET) technology is an innovative semiconductor technology that uses carbon nanotubes (CNTs) as the channel material instead of traditional silicon in the transistor design. Carbon nanotubes are cylindrical molecules made of carbon atoms arranged in a hexagonal lattice, resembling rolled- up sheets of graphene. These nanotubes exhibit remarkable electrical, thermal, and mechanical properties, making them highly attractive for electronic device applications[9].CNTFETs operate on the same principle as conventional silicon-based Field-Effect Transistors (FETs), where a gate voltage controls the flow of electrical current between the source and drain terminals through a channel. However, the use of CNTs as the channel material in CNTFETs offers several potential advantages over silicon, including higher electron mobility, greater current density, and the possibility of ballistic transport, where electrons move through the channel with minimal scattering, thus reducing power consumption and improving device performance[10].

IOP Conf. Series: Materials Science and Engineering **1316** (2024) 012005 doi:10.1088/1757-899X/1316/1/012005

The promise of CNTFET technology lies in its potential to enable the continued miniaturization of electronic components while overcoming some of the physical limitations faced by silicon-based devices, especially as the industry approaches the limits of Moore's Law. CNTFETs could lead to the development of faster, more energy-efficient electronic devices, ranging from processors and memory to sensors and flexible electronics[12]

3.1. Device parameters of CNTFET

In the detailed analysis of Carbon Nanotube Field-Effect Transistors (CNTFETs), several device parameters play pivotal roles in dictating the performance and applicability of these advanced semiconductor devices. A critical aspect of CNTFETs is their reliance on the unique properties of carbon nanotubes (CNTs), which serve as the conductive channel between the transistor's source and drain. The chirality of the CNT, denoted as (n,m), is fundamental in determining the electrical behavior of the CNT, categorizing it as either metallic or semiconducting. This distinction is vital for the design of CNTFETs, as semiconducting CNTs are preferred for transistor applications due to their ability to switch between conducting and insulating states. The diameter of the CNT, denoted as dCNT, is another crucial parameter that influences the device's electrical characteristics[13]

The diameter, which is intrinsically linked to the CNT's chirality, also affects the bandgap of the CNT, thus impacting the device's performance in semiconductor applications. Typically, dCNT values are in the range of 1-2 nm, aligning with the need for semiconducting behavior in CNTFETs. Gate length (Lg) is paramount in controlling the short-channel effects that are critical when scaling down devices for high-density applications[14].

The gate oxide thickness (tox) directly impacts the gate capacitance and, consequently, the switching speed of the device. A thinner gate oxide is beneficial for achieving higher drive currents due to the stronger capacitive coupling between the gate and the channel. However, this advantage comes at the cost of increased gate leakage, posing a design challenge. Gate oxides are often made of high-k materials to mitigate these issues, with thicknesses typically around 1-2 nm. Channel length (Lch), the distance between the source and drain contacts along the CNT, is a defining feature of CNTFETs that affects their operational speed. Reducing the channel length can potentially improve the device's speed by shortening the path electrons must travel. However, very short channel lengths may introduce undesirable short-channel effects, demanding careful optimization of this parameter. Lastly, the source/drain contact resistance (Rsd) is a significant factor affecting the overall performance of CNTFETs[15].

Table 1. Device Parameters of CNTFET

Device Parameter	Values
Chirality of tube	(19,0)
CNT diameter	0.7 nm-2 nm
Gate length	32 nm
Gate oxide thickness	1 nm- 2nm
Channel Length	32 nm

In sum, the design and optimization of CNTFETs involve a complex interplay of these parameters, each contributing to the device's final performance characteristics. The manipulation of chirality, careful selection of CNT diameter, precise control over gate length and oxide thickness, optimization of channel length, and reduction of contact resistance are all critical for harnessing the full potential of CNTFET technology in next-generation electronic devices.

IOP Conf. Series: Materials Science and Engineering **1316** (2024) 012005 doi:10.1088/1757-899X/1316/1/012005

3.2. Device performance of CNTFET

CNTFETs harness the unique properties of carbon nanotubes (CNTs) to achieve superior performance, which is characterized by several key factors. Like traditional Field-Effect Transistors (FETs), CNTFETs control the flow of current through a semiconductor channel with an electric field introduced by a gate electrode. The semiconductor in this case is a carbon nanotube, which can be either a single-walled carbon nanotube (SWCNT) or a multi-walled carbon nanotube (MWCNT), depending on the design. The source and drain contacts are made on either end of the CNT, and the gate electrode can be placed either below the CNT (as in a back-gated configuration) or surrounding it (in a top-gated configuration). The gate electrode is separated from the CNT by a thin insulating layer known as the gate dielectric. Applying a voltage to the gate electrode creates an electric field across the dielectric, and influencing the energy bands of the CNT and thereby controlling the flow of the electrons between the source and drain contacts. For n-type operation, applying a positive gate voltage attracts electrons into the CNT, lowering the energy barrier for electron transport and allowing current to flow. Conversely, when the gate voltage is removed or made negative, the energy barrier increases, preventing electron flow and turning the device off. The operation is similar but with opposite polarities for p-type CNTFETs. The quantum mechanical properties of CNTs play a crucial role in their operation as transistors. The bandgap of a CNT — the energy difference between its valence and conduction bands — determines its electrical conductivity and is directly influenced by the nanotube's diameter and chirality. This bandgap gives CNTFETs their semiconducting properties and allows for the fine-tuning of device characteristics by selecting CNTs with specific physical properties. Electron transport in CNTs is also significantly affected by quantum confinement, as electrons in the quasi-one-dimensional structure of a CNT behave as quantum waves. This confinement leads to discrete energy levels within the CNT and allows for ballistic transport of electrons, where they can travel through the CNT without scattering, under certain conditions[15].

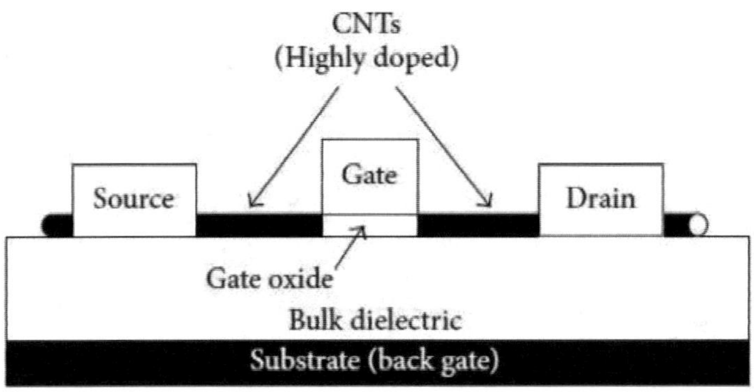

Figure 3. CNTFET diagram

4. Simulation results of existing system

The power consumption, delay, and Power-Delay Product(PDP) were observed to decrease in the CNTFET 32 nm technology utilizing when compared to 32 nm and 130 nm MOSFET technology . Further reductions in power, delay, and PDP were achieved in the proposed system, by modifying the existing system with leakage power reduction technique to Carbon Nanotube Field-Effect Transistor (CNTFET) technology. The below figure shows the output waveforms of LCG using 130nm MOSFET technology and 32nm CNTFET and MOSFET technology.

The below table presents a comprehensive performance analysis of Linear Congruential Generators (LCG) utilizing MOSFET at both 32 nm and 130 nm technology, and Carbon Nanotube Field-Effect Transistors (CNTFET) at the 32 nm technology . This analysis covers critical performance metrics such as power consumption, delay, and Power Delay Product (PDP), across several key digital components: adders, D Flip-Flops (DFF), multiplexers (MUX), and the LCG itself. Through this detailed

IOP Conf. Series: Materials Science and Engineering **1316** (2024) 012005 doi:10.1088/1757-899X/1316/1/012005

examination, comparisons are drawn to highlight the efficiency and effectiveness of the existing system across different configurations and technology.

Figure 4. Simulation result of LCG in 130nm MOSFET technology

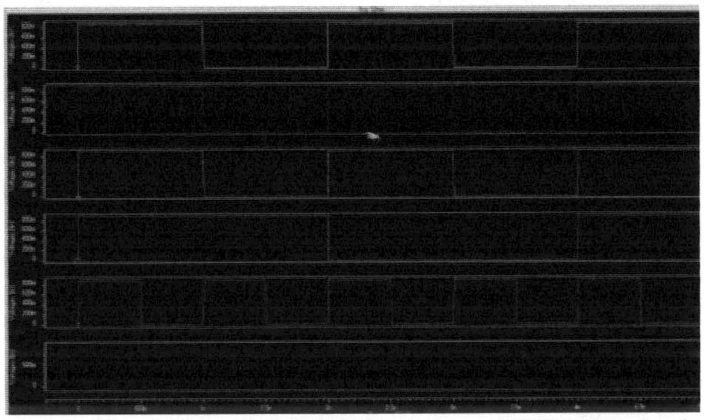

Figure 5. Simulation result of LCG in 32nm MOSFET technology

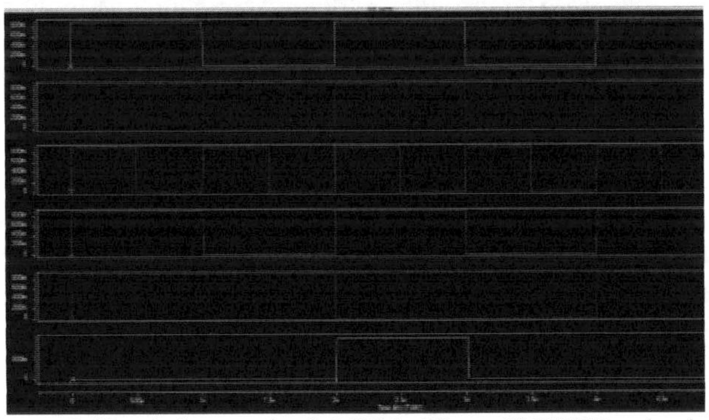

Figure 6. Simulation result of LCG in 32nm CNTFET technology

Table 2. Performance Analysis of LCG using130nm and 32nm MOSFET Technology

DEVICE	130nm MOSFET			32nm MOSFET		
	Power(W)	Delay(s)	PDP(J)	Power(W)	Delay(s)	PDP(J)
ADDER	1.715x10^-5	270.99x10^-12	4.6474x10^-15	1.697x10^-7	2.417 x10^-12	0.4103x10^-18
AND	1.790x10^-5	46.594x10^-12	0.8340x10^-15	1.962 x10^-7	4.6525x10^-12	0.9123x10^-18
DFF	4.432x10^-5	36.998x10^-12	1.6397x10^-15	0.480 x10^-6	6.1072x10^-12	2.9351x10^-18
MUX	1.752x10^-5	1.113x10^-12	0.0195x10^-15	8.33 x10^-6	152.36x10^-12	12.696x10^-18
LCG	2.096x10^-4	11.93x10^-12	2.5005x10^-15	2.058x10^-5	8.7748x10^-12	180.58x10^-18

Table 3. Comparison analysis of LCG under 32 nm CNTFET

DEVICE	**32nm CNTFET**		
	Power(W)	**Delay(s)**	**PDP(J)**
ADDER	9.471 x10^-8	498.85x10^-15	47.246x10^-18
AND	7.632 x 10^-8	57.543x10^-15	0.00439x10^-18
DFF	2.315x10^-20	57.024x10^-15	0.00132x10^-18
MUX	6.976x10^-8	27.578x10^-15	0.00192x10^-18
LCG	2.848x10^-6	32.73x10^-15	0.01061x10^-18

5. Sleep transistor technique for leakage power reduction in LCG

The sleep technique is a cutting-edge solution designed to tackle the issue of leakage power in digital circuits, a significant concern as technology continues to scale down. This method is particularly beneficial in applications requiring constant operation, like Linear Congruential Generators (LCGs), where power efficiency is critical. At the heart of the sleep technique lies its unique approach to managing power at the component level. The strategy involves the use of two transistors: a PMOS transistor connected to the positive power supply (VDD) and an NMOS transistor connected to the ground. These transistors act as switches that control the flow of power to the circuit's components. In periods of inactivity, the PMOS transistor is turned off, cutting off the power supply, while the NMOS transistor is turned on, connecting the component to the ground. This state significantly reduces the leakage current, which is a primary source of power wastage in idle digital circuits. The operational principle of the sleep technique is based on dynamically toggling the state of the circuit components between active and sleep modes, depending on their usage.

In the sleep mode, by disconnecting the components from VDD and grounding them, the technique effectively minimizes the leakage current. This is crucial for maintaining energy efficiency, especially in devices where the components are not always needed but cannot be completely powered down due to system requirements. Implementing the sleep technique across our project's key components, including D Flip-Flops (DFFs), multiplexers (MUX), and AND/OR gates, was a strategic move to combat leakage power, crucial for enhancing the energy efficiency of our Linear Congruential Generator (LCG) system. This was achieved by adding a PMOS transistor connected to the power supply (VDD) and an NMOS transistor connected to the ground for each component. These transistors serve as switches that control the power flow: when a component is idle, the PMOS transistor is turned off to cut the power supply, and the NMOS transistor is turned on to ground the component, significantly reducing leakage power. In practice, this meant integrating the sleep technique at the circuit level, ensuring each DFF, MUX, and logic gate could enter a low-power state without affecting the system's functionality.

For DFFs, which are crucial for timing and data storage, implementing sleep mode ensured they only consumed power when updating or holding data. For MUXes, which direct data paths in the LCG, and AND/OR gates, crucial for logical operations, the sleep technique minimized power usage when their operations were not required, thereby reducing overall energy consumption without compromising the accuracy or speed of the pseudorandom number generation. This approach not only showcased our project's commitment to energy efficiency but also demonstrated the practicality of the sleep technique in reducing leakage power across various digital components, ensuring our LCG system remains both effective and environmentally conscious as shown in Table 4 and 5.

6. Simulation results and comparative analysis of proposed system

After implementing the sleep technique aimed at reducing leakage power, our comparative simulation study underscored the efficacy of this approach across various digital components within Linear Congruential Generator (LCG) systems. Utilizing 32 nm Carbon Nanotube Field-Effect Transistor

(CNTFET) technology, we observed a marked enhancement in performance over the conventional 130 nm and 32 nm MOSFET technology. This improvement was consistently evident in digital components such as adders, AND gates, D Flip-Flops (DFF), multiplexers (MUX), and the LCG itself, where key metrics like power consumption, delay, and Power Delay Product (PDP) saw significant advancements. The simulation results, alongside detailed tables, comprehensively illustrate the beneficial impact of incorporating sleep techniques with CNTFET technology, highlighting its potential in advancing the energy efficiency of LCG systems and potentially other digital circuits.

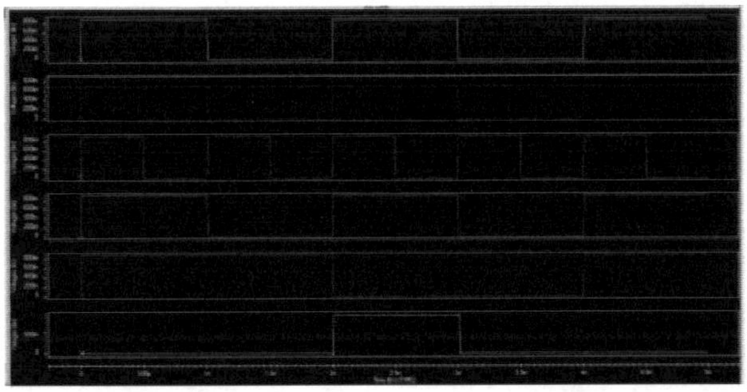

Figure 7. LCG with sleep technique using 130nm MOSFET technology

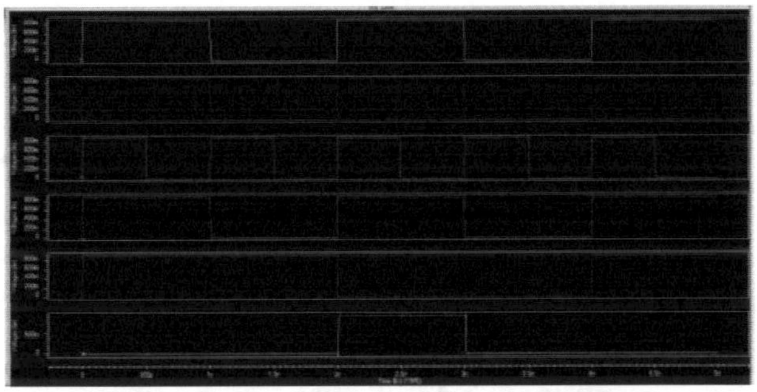

Figure 8. LCG with sleep technique using 32nm MOSFET Technology

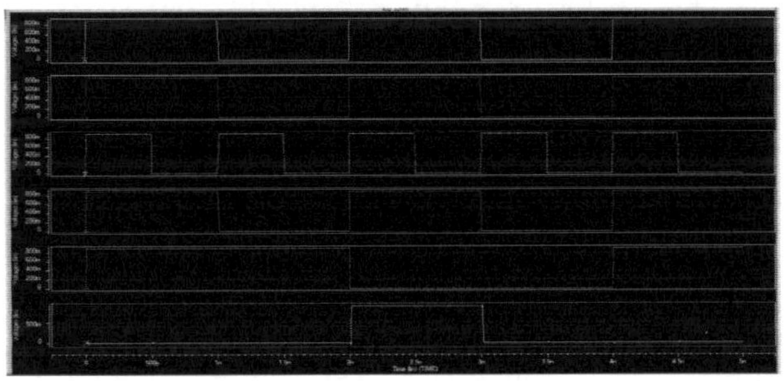

Figure 9. LCG with sleep technique using 32nm CNTFET Technology

Table 4. Performance Analysis of LCG with sleep technique using 130nm and 32nm MOSFET Technology

DEVICE	130nm MOSFET			32nm MOSFET		
	Power(W)	Delay(s)	PDP(J)	Power(W)	Delay(s)	PDP(J)
LCG	2.096x10^-4	11.93x10^-12	2.5005x10^-15	2.058x10^-5	8.7748x10^-12	180.58x10^-18
LCG with sleep technique	3.286 x10^-5	9.27x10^-12	0.0874x10^-15	2.684 x 10^-7	5.569x10^-12	16.024x10^-18

Table 5. Comparison analysis of LCG with sleep technique using 32nm CNTFET Technology

DEVICE	32nm CNTFET		
	Power(W)	**Delay(s)**	**PDP(J)**
LCG	2.848x10^-6	32.73x10^-15	0.01061x10^-18
LCG with sleep	5.144 x 10^-17	5.7579x10^-15	0.00024x10^-18

7. Conclusion

The integration of sleep techniques for leakage power reduction within Linear Congruential Generators (LCG), employing 32 nm Carbon Nanotube Field-Effect Transistor (CNTFET) technology, marks a significant advancement over conventional 130 nm MOSFET approaches. Through detailed comparative simulations, it became evident that CNTFET technology, coupled with sleep techniques, offers substantial improvements in power efficiency. This is demonstrated by notable reductions in power consumption, delay, and optimized Power Delay Product (PDP) across a range of digital components. These findings not only highlight the potential of CNTFET technology in enhancing energy efficiency but also suggest a promising direction for future research in digital circuit optimization. This project contributes valuable insights into the development of more sustainable electronic devices, emphasizing the importance of innovative approaches in semiconductor technology.

8. References

[1] C. Shravan, S. K. Marri and T. Kampasati, "VLSI Architectures of Three Operand Binary Adders," 2023 International Conference on Advances in Electronics, Communication, Computing and Intelligent Information Systems (ICAECIS), Bangalore, India, 2023, pp. 338-343, doi: 10.1109/ICAECIS58353.2023.10170061.

[2] A. Kumar Panda and K. Chandra Ray, "A Coupled Variable Input LCG Method and its VLSI Architecture for Pseudorandom Bit Generation," in IEEE Transactions on Instrumentation and Measurement, vol. 69, no. 4, pp. 1011-1019, April 2020, doi: 10.1109/TIM.2019.2909248.

[3] K. S. T. Ramapragada, A. K. R. Netla, P. K. Chattada and B. Manickam, "Design and FPGA Implementation of High-Speed Area and Power Efficient 64-bit Modified Dual CLCG based Pseudo Random Bit Generator," 2021 IEEE International Symposium on Smart Electronic Systems (iSES), Jaipur, India, 2021, pp. 93-98, doi: 10.1109/iSES52644.2021.00032.

[4] Sathanur, Ashoka, et al. "Optimal sleep transistor synthesis under timing and area constraints." Proceedings of the 18th ACM Great Lakes symposium on VLSI. 2008.

[5] Kim, Se Hun, and Vincent J. Mooney. "Sleepy keeper: a new approach to low-leakage power VLSI design." 2006 IFIP International Conference on Very Large Scale Integration. IEEE, 2006.

[6] Sathanur, Ashoka, et al. "Row-based power-gating: A novel sleep transistor insertion methodology for leakage power optimization in nanometer CMOS circuits." IEEE Transactions on Very Large Scale Integration (VLSI) Systems 19.3 (2009): 469-482.

[7] Island, Generic Voltage, and C. A. D. Flow. "One of the major techniques used to control subthreshold is sleep transistors. In essence, sleep transistors are used for power gating: Logic runs at low Vt, and the gates are faster and leaky; Sleep transistors are high Vt. They are switched off when idle (usually NMOS alone is used) and can save 2-1000× leakage power. Your goal is to design a 32 kbit SRAM (128 rows, 256 columns, 8 bit words) which uses." Solid State Circuits (2002): 1421-1432.

[8] Sujitha, A., et al. "Analysis of an Efficient Linear Congruential Generator Architecture for Digital Applications." 2023 Fifth International Conference on Electrical, Computer and Communication Technologies (ICECCT). IEEE, 2023.

[9] Tezuka, Shu. "Linear congruential generators." Uniform Random Numbers: Theory and Practice. Boston, MA: Springer US, 1995. 57-82.
[10] L'ecuyer, Pierre. "Tables of linear congruential generators of different sizes and good lattice structure." Mathematics of Computation 68.225 (1999): 249-260.
[11] Marsaglia, George. "The structure of linear congruential sequences." Applications of number theory to numerical analysis. Academic Press, 1972. 249-285.
[12]Kurlberg, Pär, and Carl Pomerance. "On the period of the linear congruential and power generators." arXiv preprint math/0405120 (2004).
[13] Khan, Imran Ahmed, et al. "Design of low-power CNTFET parity generators for high-speed data transmission." Innovations in Electronics and Communication Engineering: Proceedings of the 9th ICIECE 2021. Singapore: Springer Singapore, 2022. 605-611.
[14] A. K. Panda, R. Palisetty and K. C. Ray, "Area-Efficient Parallel-Prefix Binary Comparator," 2019 IEEE International Symposium on Smart Electronic Systems (iSES) (Formerly iNiS), Rourkela, India, 2019, pp. 12-16, doi: 10.1109/iSES47678.2019.00016.
[15] A. Thane and R. Chaudhari, "Hardware Design and Implementation of Pseudorandom Number Generator Using Piecewise Linear Chaotic Map," 2018 International Conference on Advances in Computing, Communications and Informatics (ICACCI), Bangalore, India, 2018, pp. 456-459

IOP Conf. Series: Materials Science and Engineering **1316** (2024) 012006 doi:10.1088/1757-899X/1316/1/012006

Design of 8T DTMOS Schmitt Trigger SRAM Cell for IOT Applications

Neeta Pandey[1*], Aditya Singh Mann[2] and Abhay Setia[2]

[1] Department of Electronics and Communication Engineering, Delhi Technological University, New Delhi, India
[2] Department of Electronics and Communication Engineering, Delhi Technological University, New Delhi, India

*E-mail: neetapandey@dce.ac.in

Abstract. Internet of things (IoT) based systems require power-efficient circuits to raise the battery lifeline. This study presents a single-ended 8T SRAM cell. The core of the proposed 8T SRAM cell is composed of a Schmitt-Trigger circuit which a dynamic body bias technique is applied to a standard CMOS inverter through a feedback mechanism, whereby the threshold voltages of two MOSFETs can be changed, thus changing the switching voltage. Read operation of the proposed cell is conducted using the shared footer per word transistor. The write path is cut-off during the read operation, improving RSNM. A transmission gate placed in the cell core is used to cut the feedback path during write operation. To prove superiority of the proposed cell it is compared with four state-of-the-art SRAM cells under identical conditions on Cadence Virtuoso using 45nm technology at VDD=0.8 V. The proposed circuit shows a 135.74 % improvement in terms of RSNM and a 44.04 % improvement in terms of peak-to-peak power compared to the 6T DTMOS Cell.

1. Introduction

In the realm of modern integrated circuits, Static Random Access Memory (SRAM) stands as a cornerstone, crucial for various computing systems and applications. Its significance is particularly pronounced in the Internet of Things (IoT) landscape, where power efficiency is paramount. As IoT devices proliferate, spanning sensors, actuators, and wearables, the demand for power-efficient memory solutions grows [1].

SRAM finds diverse applications in IoT, including data caching, buffering, and temporary storage of sensor readings. Designing SRAM cells tailored for IoT is crucial, considering factors like read stability [2]. This paper presents a novel 8T SRAM cell design integrating a Schmitt-Trigger circuit with dynamic body biasing, aimed at reducing power consumption while ensuring reliable data storage and retrieval. Through comparative analysis, the proposed design demonstrates superior performance metrics compared to existing solutions, highlighting its potential to enhance energy efficiency and prolong battery life in IoT devices. Innovative memory design methodologies like this are instrumental in meeting the evolving demands of IoT systems, fostering seamless integration into the expanding IoT landscape.

IOP Conf. Series: Materials Science and Engineering **1316** (2024) 012006 doi:10.1088/1757-899X/1316/1/012006

Further, Dynamic Threshold Metal-Oxide-Semiconductor (DTMOS) is used in different applications including SRAM, to enhance performance and mitigate power consumption challenges. Its implementation in SRAM design offers several advantages that contribute to better efficiency and reliability [3]. DTMOS technology presents a promising avenue for SRAM design, offering dynamic threshold voltage control that can minimize leakage currents during standby modes, performance, and robustness against variations. Its implementation in SRAM architecture aligns with the growing demand for energy-efficient and adaptable memory solutions across various computing and electronic applications [4].

In this paper, an 8T SRAM cell is put forward that uses a Schmitt-Trigger circuit wherein dynamic body bias technique is applied. The read operation of the proposed cell employs a shared footer per word transistor. A transmission gate placed in the cell core is used to cut the feedback path during write operation. The paper is arranged in six sections. Literature review is presented in Section II followed by description of performance parameters in Section III. The proposed cell is described in Section IV. The simulation results are presented in Section V and the paper concludes in Section VI.

2. Literature Review

In order to make innovative interventions more effective, we need to improve memory systems. To do this, different designs have been suggested, ranging from 6T to 18T SRAM cells. These designs include various ways of reading and writing data, like single-end or differential methods. Researchers have also looked into ways to use less power and make access faster, without sacrificing other important aspects of performance. The following section aims to summarize and analyze the research done in this area, to see what methods work best for improving memory systems.

Erfan Abbasian et al.[5] proposed a novel 7T Single Ended Schmitt Trigger based SRAM that uses DTMOS. In this cell the read decoupling technique is used to improve RSNM and to enhance WSNM, there is a feedback-cutting mechanism. It demonstrates extremely low T_{RA}. Raghav Shekhar et al.[8] proposed a 10T differential SRAM cell that uses 6T SRAM cell along with additional four transistors in the form of a vertically oriented inverter whose output is attached to the source of the NMOS of another inverter. This configuration prevents any inversion taking place during the read process by restoring the value. This cell provides large hold and read noise margins. Ravi Teja Yekula et al.[11] proposes a 13T SRAM [add technique] cell that has a shorter read access time than existing circuits and has applications for deep-space. It consists of a stacked NMOS structure consisting of three transistors which can perform a differential read operation without affecting the nodes of the SRAM cell. S. Cai et al.[12], the author's have proposed an 18T SRAM cell which is based on Muller C elements. The circuit has 4 storage nodes, namely, Q, QB, A and B. It is immune to Single Event Upset (SEU) and demonstrates high-reliability but a certain area overhead. Yinghuan Lv et al.[13] proposed a radiation hardened 8T structure, which utilizes the concept of cutting the inverter feedback, with the help of NMOS transistors attached to the path between the drain of the PMOS and the drain of the NMOS.It has large read current, low leakage current, and a large RSNM with some area sacrifice.

3. SRAM Cell Performance

One of the fundamental building blocks of memory architectures is the 6T SRAM cell, that provides high-speed and reliable data storage in electronic systems. Comprising six transistors arranged in a specific configuration, the 6T SRAM cell operates based on bistability principles, allowing it to retain data until power is provided to the system. A six-transistor (6T) SRAM bitcell serves fundamentally as a unit within the SRAM array as depicted in Fig 1 (a). The 6T SRAM bitcell comprises two access transistors (M1 and M2) and two back to back connected inverters (M3-M4, M5-M6) linked to each data storage node. The paired inverters create a latch that has binary information, storing the data's true and complementary versions. Figure 1(b) shows the DTMOS based counterpart of 6T SRAM bit cell.

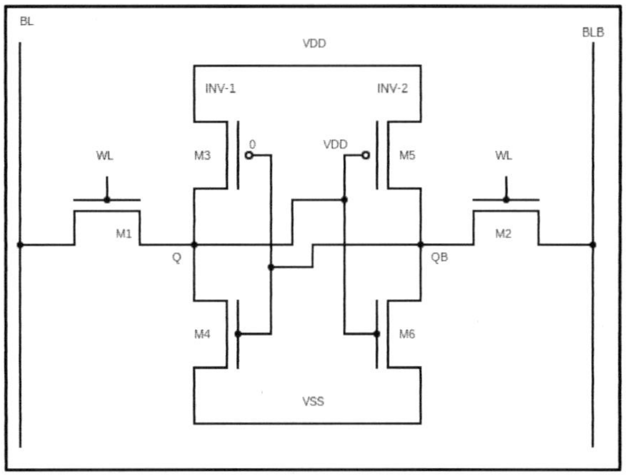

Figure 1(a). 6T SRAM Bitcell

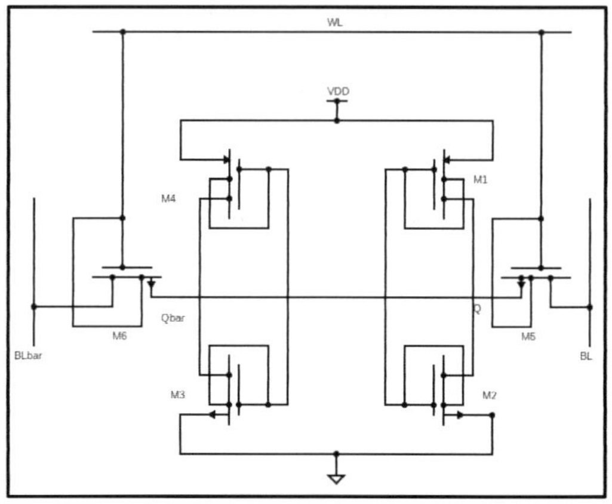

Figure 1(b). 6T SRAM Bitcell based on DTMOS

Noise margin, Read and Write access time; and Power are some of the important parameters to study the performance of an SRAM. These parameters are described in the following text:

Noise Margin:
The prime indicator in an SRAM bitcell used to gauge its stability during read cycles and in a hold state is the Static Noise Margin (SNM). It represents the maximum level of DC noise (VN) that the cross-coupled inverter pair can withstand while retaining the bitcell's data. Extracting the read SNM involves observing the read voltage transfer characteristics (VTC). From Fig 1(a), this measurement entails checking the voltage across the data storage node Q (or QB) while maintaining bias at VDD for both bitlines (BL, BLB) and the wordline (WL), while the node voltage at QB (or Q) is concurrently monitored. Read and write access times are crucial parameters in the operation of 6T SRAM, determining how quickly data can be retrieved from or written into the memory cell.

Read and Write Access Time:
The read access time(TRA) in a 6T SRAM cell refers to the duration required to access stored information from the memory cell. The read access time is affected by various factors such as Cell Size, Sense Amplifier Speed and Bitline Capacitance. During a read operation, the stored data is accessed by sensing the voltage difference between the bitlines. The write access time(TWA) refers to the duration necessary to write new data into the SRAM cell. Factors influencing write access time include Write Driver Speed, Precharge Time and Cell Size This process involves modifying the cell's state to store new information.

Power:
Power is measured while performing the DC operating point analysis, that is, it is measured when we are sweeping the node QB and the bitlines and the worldline are set to VDD. The power is obtained using the inbuilt analysis feature of Virtuoso. Several parameters, including power dissipation, are optimized to achieve better efficiency in SRAM designs.

IOP Conf. Series: Materials Science and Engineering **1316** (2024) 012006 doi:10.1088/1757-899X/1316/1/012006

4. Proposed 8T DTMOS Cell

The proposed cell employs the cascaded inverter based DTMOS Schmitt Trigger [14].The schematic of the proposed circuit is given in Fig. 2. In the circuit NM1-PM0 and PM1-NM0 form two inverters, such that the transistors' body of one is connected to the output of the other. An inbuilt body bias is provided through the dynamic threshold configuration which forms the basis of our feedback mechanism. The proposed bitcell uses a single bitline(BL) to form read and write operations. A feedback cutting transmission gate has been placed inside the cell core which is controlled by the WWL and WWLB signals. The following subsection explains the working of the proposed cell.

Read Operation:
The read operation is helped by the read assist transistors and provides a discharging path by utilising the opposite nature of the WWLB signal. The write path is cutoff by pulling down the WWL signal. The read assist transistors NM4 and NM5, gated by the node QB and WWLB signal respectively and help in the discharging process. The NM5 read assist transistor is shared per word.

Write Operation:
The write operation is initiated by the pulling up the WWL signal, which cuts the feedback mechanism and establishes the write path; now the circuit behaves as two cascaded inverters. In order to write '1' at storage node 'Q', the BL is set to VDD which drives the circuit to its complement value.

Hold Operation:
During the hold operation the WWL is pulled down in order to establish the feedback path and the bitline is pre-charged to VDD and the inverters form a cross-coupled pair.

5. Simulation

The simulation has been carried out using the Cadence Analog Design environment, in order to verify the read, write and hold operations with a supply voltage of 0.8 V. The circuit has been tested for a timespan of 45 ns in which from 0 to 15 ns, a value is written into the node 'Q' to initiate charge reversal. From 15 to 30 ns, we are holding the value, and from 30 to 45 ns we are reading the storage node. The key timing signals here are WWL and WWLB which initiate the operations by cutting and establishing the feedback path during various stages.

The noise margin has been obtained by keeping in mind the complementary nature of various signals, as during the read operation we do not want to activate the write path, in Fig. 4, we have provided the butterfly curve obtained for SNM. The Write SNM is calculated by obtaining the Write and Read VTCs and plotting the smallest square possible between the two curves. The power has been obtained using the inbuilt peak-to-peak power feature of Virtuoso using the calculator.

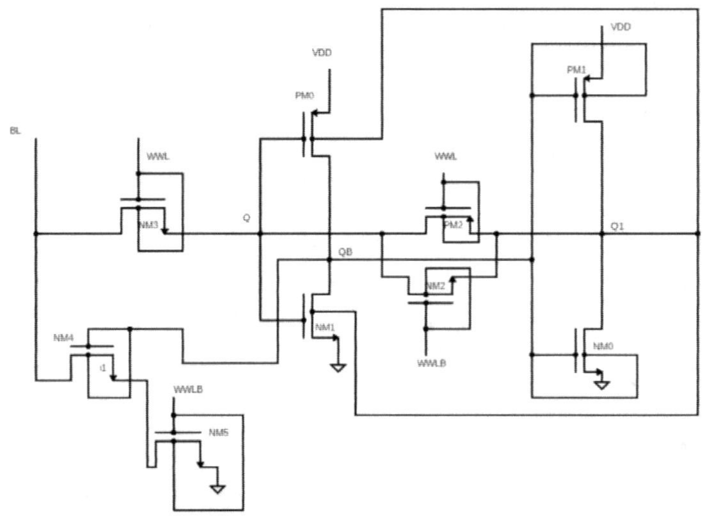

Figure 2. 8T DTMOS (Proposed)

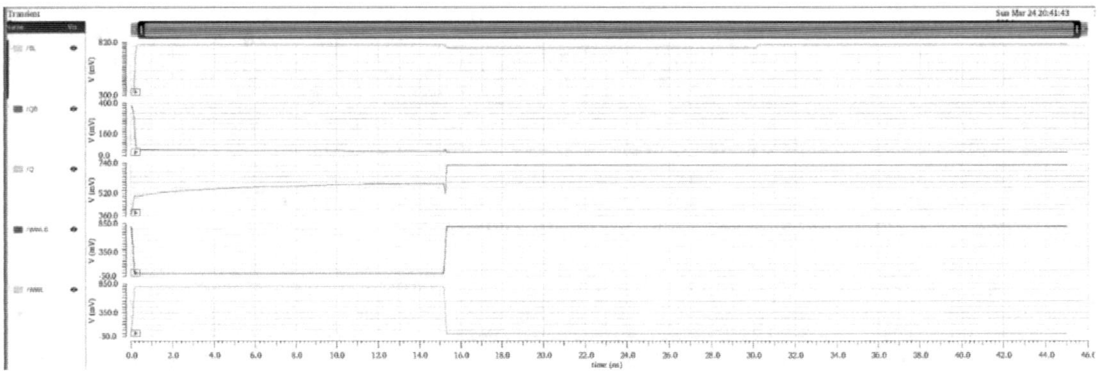

Figure 3. Read and Write Timing Diagram of Proposed 8T DTMOS SRAM Bitcell

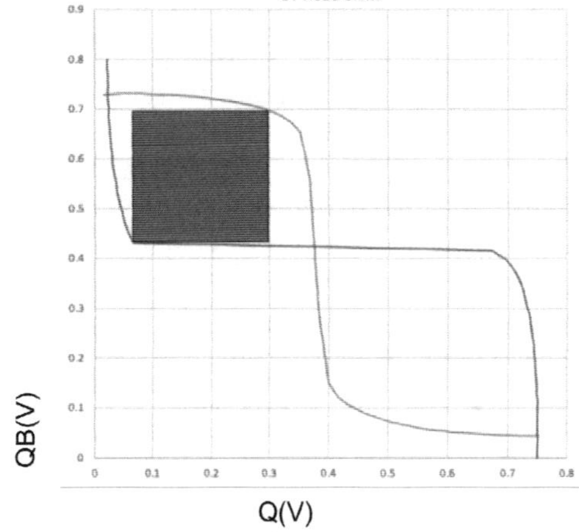

Figure 4. Read SNM of Proposed 8T DTMOS SRAM Bitcell

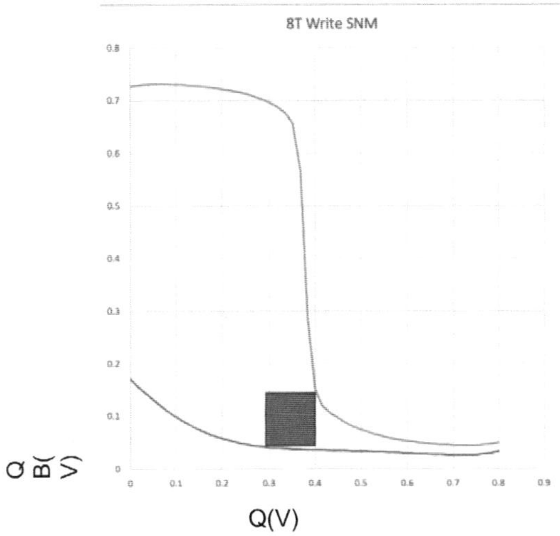

Figure 5. Write SNM of Proposed 8T DTMOS SRAM Bitcell

Table 1. Parameter Comparison for Various Designs

Parameters	**DTMOS 6T**	**Schmitt Trigger 7T [5]**	**RHD 8T [13]**	**RHD 10T [8]**	**8T DTMOS (Proposed)**
Read SNM (V)	0.09771	0.210504	0.09951	0.13544	0.23035
Write SNM (V)	0.09961	-	0.16035	0.05327	0.0957
TRA (ps)	50	60.678	71.7	85	76.1
TWA(ps)	327.68	207.1831	144.19	192.22	84
Power (Peak to Peak) (uW)	72.2	129	90.4	61.6	40.4

6. Conclusion

In this study, we proposed a novel 8T DTMOS SRAM Cell, which employs the Schmitt Trigger topology. The proposed shows the best RSNM among all the bitcells studied and also demonstrates a good TRA and TWA. It improves upon the existing topologies by employing the transmission gate technique and does away with the read decoupling technique. The key variable that we wanted to optimize was power and a major reduction has been observed after performing peak-to-peak power analysis. The power has reduced from 72.2 uW in the 6T DTMOS SRAM Cell to 40.4 uW in the proposed SRAM cell, marking a 44.04 % reduction. Thus, in comparison to the circuits study the proposed circuits stands out with regard to Power and Read SNM in the context of 45nm technology.The proposed circuit is first in terms of RSNM, TWA and peak-to-peak power and third in terms of TRA. Since we have successfully managed to reduce the power consumption of SRAM bitcell which makes it suitable for IoT applications in which energy efficient bitcells are necessary to raise the battery lifetime.

References

[1] A. Banerjee, S. H. Dhong, M. C. McElroy, "Designing energy-efficient SRAM cells for IoT applications," 2018 IEEE Custom Integrated Circuits Conference (CICC), San Diego, CA, USA, 2018, pp. 1-4.

[2] R. K. Dokania, R. Saravanan, "Design and optimization of SRAM for IoT applications," 2016 International Conference on Microelectronics, Computing & Communication Systems (MicroCom), Durgapur, India, 2016, pp. 1-6.
[3] H. C. Huang, C. H. Lin, "Low-power and high-speed SRAM design using DTMOSFETs for IoT applications," 2017 IEEE International Symposium on VLSI Design, Automation and Test (VLSI-DAT), Hsinchu, Taiwan, 2017, pp. 1-4.
[4] L. Zhang, Y. Chen, "Advanced SRAM architectures for power-efficient IoT devices," 2020 IEEE International Conference on Integrated Circuits and Microsystems (ICICM), Beijing, China, 2020, pp. 1-4.
[5] Abbasian, E., & Gholipour, M. (2021b). Design of a Schmitt-Trigger-Based 7T SRAM cell for variation resilient Low-Energy consumption and reliable internet of things applications. AEU - International Journal of Electronics and Communications, 138, 153899. https://doi.org/10.1016/j.aeue.2021.153899
[6] Panchal, A., Sharma, P., Gupta, A., Neema, V., Tiwari, N., & Sindal, R. (2023). Improved reliability single loop single feed 7T SRAM cell for biomedical applications. Memories - Materials Devices Circuits and Systems, 4, 100057. https://doi.org/10.1016/j.memori.2023.100057
[7] Roy, C., & Islam, A. (2021). Design of low power, variation tolerant single bitline 9T SRAM cell in 16-nm technology in subthreshold region. Microelectronics Reliability, 120, 114126. https://doi.org/10.1016/j.microrel.2021.114126
[8] Shekhar, R., & Kumar, C. I. (2022). Design of highly reliable radiation hardened 10T SRAM cell for low voltage applications. Integration, 87, 176–181. https://doi.org/10.1016/j.vlsi.2022.07.004
[9] Mittal, Sparsh & Verma, Gaurav & Kaushik, Brajesh Kumar & Khanday, Farooq. (2021). A Survey of SRAM-based In-Memory Computing Techniques and Applications. Journal of Systems Architecture. 119. 10.1016/j.sysarc.2021.102276.
[10] Sugii, N. (2015). Low-power-consumption fully depleted silicon-on-insulator technology. Microelectronic Engineering, 132, 226–235. https://doi.org/10.1016/j.mee.2014.08.004
[11] Teja Yekula R., Pandey M. & Islam A., "A Highly Reliable Radiation Tolerant 13T SRAM Cell for Deep Space Applications," in Microelectronics reliability, vol. 1330, pp. 1–14, Jun. 2022. Publisher: Elsevier, DOI: https://doi.org/10.1016/j.microrel.2022.114527
[12] Cai, S., Wen, Y., Ouyang, J., Wang, W., Yu, F., & Li, B. (2023). A highly reliable and low-power cross-coupled 18T SRAM cell. Microelectronics Journal, 134, 105729. https://doi.org/10.1016/j.mejo.2023.105729
[13] Lv, Y., Wang, Q., Ge, H., Xie, T., & Chen, J. (2021). A highly reliable radiation hardened 8T SRAM cell design. Microelectronics Reliability, 125, 114376. https://doi.org/10.1016/j.microrel.2021.114376
[14] Zhang C, Srivastava A, Ajmera P. Low voltage CMOS Schmitt trigger circuits. Electron Lett 2003;39:1696–8.

The effect of Dolomite powder and Fly ash on durability properties of self-compacting concrete

R. Vaishnava Kumar*, A. Prabaghar

Civil and Structural Engineering Department, Annamalai University, Chidambaram, India.

apurupvishnu@gmail.com, aprabaghar@gmail.com.

Abstract. This study aims to assess the effectiveness of dolomite mineral admixture in the production of self-compacting concrete (SCC). The characteristics of concrete at 7 and 28 days of hardening were evaluated. SCC mixtures with different amounts of dolomite powder and fly ash (FA) at a constant composition were made. It has been noted that the flow ability of SCC mixes gets better with increasing percentages of dolomite. In addition to its mechanical characteristics, the durability of mixes containing SCC has been studied, including water absorption, sorptivity, and permeability to chloride ions and acid attack. After 28 days, a strength of 61 MPa was achieved with a 20% DP replacement level mixture, which is 31.38% higher than the control SCC mix. Along with the compressive strength of SCC, parameters affecting the durability of SCC such as Rapid chloride penetration test (RCPT), Sorptivity, water absorption and hydrochloric acid attack were also studied at 28 days. The utmost resistance to permeability of chloride ions was attained.

Keywords: dolomite powder, self-compacting concrete, fly ash, sorptivity, RCPT

1. Introduction

Cement, fine and coarse aggregates, and water form the composite concrete substance. When it comes to durability, concrete is perfect because it remains well-maintained even when exposed to severe environments. There will be lesser air pollution when the fine powder wastes are used as filler materials in concrete. Because of its unique properties, self-compacting concrete (SCC) can compact itself into densely packed, narrow sections simply by using its own weight. It can also consolidate itself without the need for internal or external vibration [1, 2-4]. It can maintain its cohesiveness and stability without causing bleeding or segregation. A minor amount of superplasticizer is present in SCC. Many reports indicate that SCC contains powder in concentrations ranging from 450kg/m^3 to 600 kg/m^3 in concrete. In order to control the cement content, lower the heat of hydration, improve and maintain workability, and meet its rheological requirements, reactive and inert fillers are frequently utilized in SCC. A portion of this powder can be adequately substituted by mineral additives such as FA, LP, SF, GGBS, etc. A by-product of coal-fired power plants, fly ash is widely used in mixed cements. Worldwide coal ash production is expected to reach 700 million tons per year, of which at least 70% is FA [2, 3]. Reused FA only makes just about 4% of total FA consumption, despite being a valuable mineral component for PC and concrete mixtures. Utilizing chemical admixtures and large amounts of PC can be expensive when using SCC, which is one of its drawbacks. An alternate method to decrease the cost of self-consolidating concrete (SCC) is the utilization of mineral admixtures, such as fly ash (FA). Utilizing reuse FA can enhance the workability of the concrete mixture without incurring additional expenses [5, 6]. Utilizing flyash also decreases the need for cement, sand, and fine fillers that are necessary in self-consolidating concrete (SCC)[4]. Furthermore, the inclusion of reuse FA also prevents the necessity for viscosity enhanced chemical admixtures. The reduced moisture content of the concrete leads to increased durability, as well as enhanced mechanical integrity of the building. Utilizing fine aggregate as a substitute for regular Portland cement (PC) enhances the toughness and durability of self-compacting concrete (SCC) by generating a more compact matrix, thereby enhancing the endurance of concrete buildings [6-8]. Furthermore, the study on the use of dolomite powder (DP) as the filling component of SCC also produced some intriguing findings. As a result, the study's first findings indicated that there was a good chance of producing commercial SCC/SCM using DP's industrial by-product as one of the main filler fractions. Nevertheless, it's possible that research on the use of such DP for concrete/mortar based advanced construction materials was restricted to standard, usable concrete. [9, 18]. The DP concrete mix demonstrated superior defence against sulphate and chloride attacks. There was just a slight drop in compressive strength. [10]. From this raw material, a low-firing binder with low CO2 emissions can be produced. The usage of secondary waste (dolomite screens, CO2 emissions), low fire temperature, and rapid production cycle all contribute to lower costs for finished products [11]. Dolomite speeds up GGBS geopolymer concrete setting, decreasing construction time. Compressive strength was highest at 70:30 GGBS/dolomite. GGBS-Dolomite geopolymer concrete reached 95% compressive strength after 7 days of ambient curing, while traditional concrete took 28 days. GGBS-Dolomite GPC resists sulphate and saltwater assault and absorbs water well [12]. According to the EFNARC guideline for self-compacting concrete, SCCs with fly ash and dolomite are categorized as SF1 in terms of slump flow. Based on V-funnel flow time, these viscosities are classed as VF1. According to the rise in V-Funnel time at T5 minutes, Mixes with a 3:1 fly ash-dolomite powder ratio, met specifications similar to the EFNARC guide for which the SCC was built. Superplasticizer dosage was found to be the lowest in this blend [13, 14, and 16]. The calcined dolomite powder and Rice husk ash binary binder system results in better workability, drying shrinkage, and mechanical strength compared to regular Portland cement [15].

The previous researchers concentrated on strength properties of SCC. The presented research is focussed on durability of SCC having fly ash and dolomite. The purpose of this research is to investigate durability of SCC having fly ash and dolomite powder. The durability of SCC having flyash and dolomite powder is assessed by studying the RCPT, sorptivity, acid attack, water absorption test results. The aim of the experimental work is to use fly ash and dolomite powder to experimentally investigate the viability of durability criteria for self-compacting concrete (SCC).

2. Experimental program

The experimental study of the work on self-compacting concrete with a replacement of mineral admixtures were discussed in the below following sections in detail.

2.1 Materials

In this experimental examination, conventional Portland cement with a specific surface area of 298.65 m^2/kg and a confinement to 53 grade was utilized in accordance with the constraints specified in standard guidelines as per IS 12269. In accordance with codal guidelines Coarse aggregate of crushed stone with a maximum size of 16 mm and natural river sand for fine aggregate is used. The specific gravities of river sand and crushed stone are 2.6 and 2.74 respectively. FA and DP were also utilized in SCC. FA obtained from the Vijayawada Thermal Power Plant is eligible for ASTM C 618 Class C classification. DP was used raw, unprocessed, as fine powder in SCC. It was a by-product of the quarry crushers from the lime stone deposits. Chemical admixture like polycarboxylate ether was used to develop the flow properties and to enhance the durability by means of reducing the water.

Table 1. Physical properties of the materials.

Property	Cement	Flyash	Dolomite powder	Fine aggregate	Coarse aggregate	Superplasticizer
Specific gravity	3.15	2.3	2.86	2.6	2.74	1.08
Colour	Grey	Bright grey	White	Light brown	grey	Brown

The physical and mechanical properties of ordinary Portland cement and other constituted materials used in this experimental work are presented in Table 1 and chemical properties of flyash and cement [3] which are represented in Table 2 these limits are confined by IS 3812 and IS12269. The chemical composition of Dolomite powder is calcium carbonate 54%, silicon dioxide 2%. The microstructure of materials was tested by SEM as shown in figure 1.

Table 2 chemical properties of Fly ash and Cement [3]

Chemical analysis (%)	Fly ash	Cement
Silicon dioxide (SiO_2)	55.8	21.56
Calcium oxide (CaO)	4.12	49.88
Iron oxide(Fe_2O_3)	6.66	6.17
Magnesium oxide(MgO)	1.9	4.51
Aluminium oxide(Al_2O_3)	20.75	6.67
Sulphur trioxide (SO_3)	0.44	2.75
Potassium oxide (K_2O)	1.73	0.76
Sodium oxide (Na_2O)	0.78	0.43
Loss of ignition (LOI)	1.95	2.79

2.2 Mix design and Mix proportions

The self-compacting concrete was made using a trial-and-error technique based on European standards. Initially, few trials ended up unsatisfactory results but after following the recommended guidelines the control mix proportion with M40 grade was produced successfully. In total eight series of mix proportions one sequence of mix proportion serve as the control mix proportion, rest of seven mix proportions containing mineral admixtures were tested in different combinations. A constant proportion (18%) of fly ash was replaced in place of Cement and remaining cement was substituted for seven different proportions (10%, 20%, 30%, 40%, 50%, 60%, and 70%) with dolomite powder in DP series. Following the primary analyses, the water-powder ratio (w/p) was adopted and the whole powder content was chosen 550 kg/m^3. The next generation polycarboxylate based superplasticizer that standardised specifications of ASTMC 494 Type F was utilized in this study.

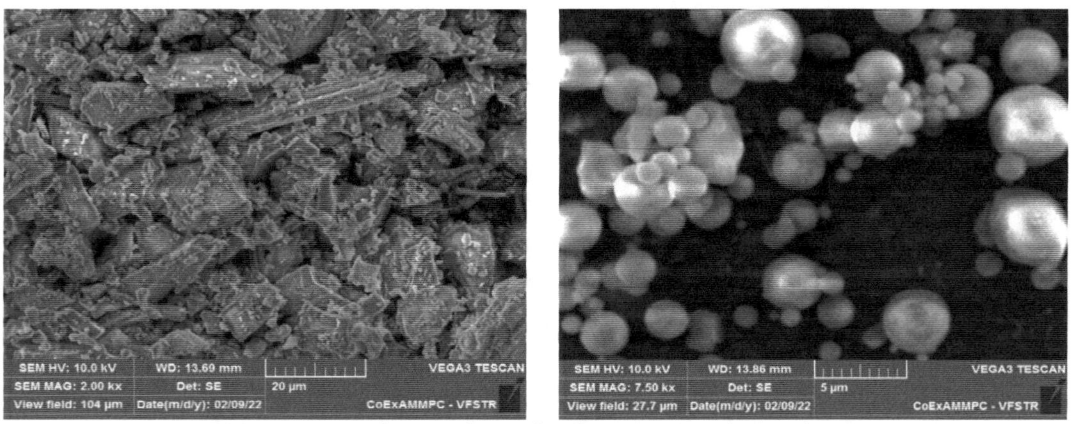

Figure 1. SEM images for Dolomite (20µm) and Fly ash (5µm).

2.3 Mixing, Sampling, Casting and curing

By implementing this approach, a very workable combination without any bleeding or segregation was obtained. To achieve uniform mixtures, it is necessary to thoroughly incorporate the cement paste with the coarse aggregate. Using this method of mixing, the aggregates in the pan mixture were originally mixed uniformly with fine material and sand for 1 minute. Then, a water content of 70% was added and mixed for an additional 1 minute. Allow the mixture to stand uninterrupted for a further minute to allow the aggregates to absorb water. Finally, the remaining 30% of water was used to include the high-end superplasticizer into the mixture, which was then mixed for an additional 3 minutes. The mixing is continued for another 5min to obtain SCC. Slump-flow, T50, L-box and V-funnel tests are to be conducted as workability tests on fresh concrete before casting [1]. Using steel moulds, specimens (cubes) are casted and precautions are to be taken that they should not undergo any compaction except their self-weights. Until demoulding the specimens were kept in a chamber at 27±2 °C for 24 h. After that, specimens are to be immersed in water at 27 °C.

3. Fresh properties

Filling, flowing, and passing abilities are the metrics used to assess the workability of SCC. The flow ability, filling capacity, and passing ability of fresh mixtures were assessed using the slump flow test, V-funnel test, and L-box test as shown in figure 2. Slump flow test instructions state that the slump cone must be filled without compaction, lifted vertically, and the concrete's two spread diameters of the resulting circle measured in two perpendicular directions and the results are classified as SF2 as per guidelines. This procedure is outlined in ASTM C1611/C1611M (ASTM 2009b) and EFNARC (2005) recommendations. The time it takes the flowing mixture to achieve a 500 mm diameter during the slump flow test is also known as the T50 time. The V-funnel flow time was measured by emptying the concrete in funnel. The blocking ratio was calculated from the heights of concrete which is taken at the time of testing. Several sources suggest different ways for testing their self-compatibility parameters, all parameters results are comes under the limits of codal guidelines.

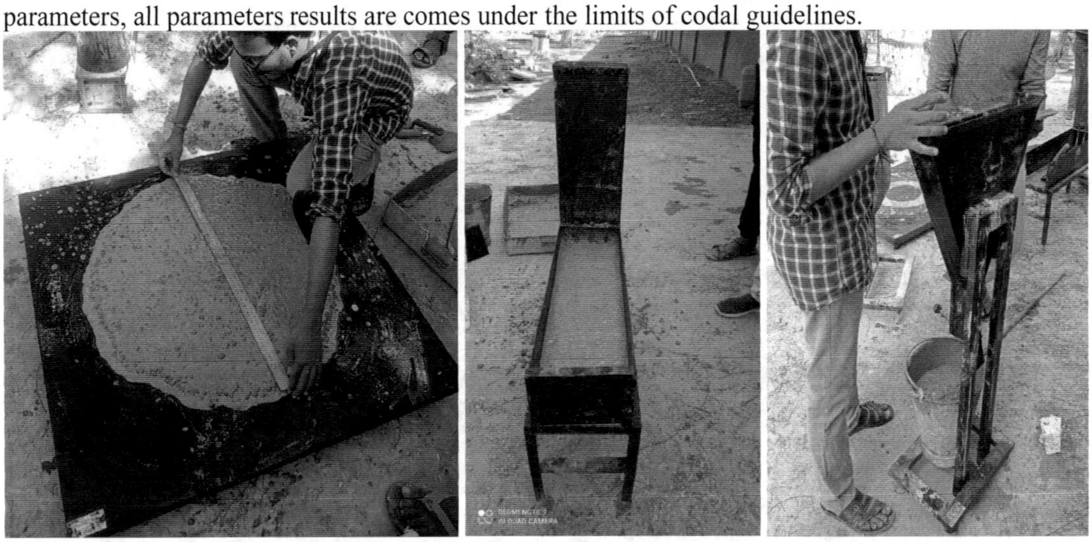

Figure 2 fresh properties of concrete slump flow test, L-box test, and V-Funnel test.

4. Properties of hardened concrete

The ultrasonic pulse velocity, density, rebound hammer, and compressive strength were measured using cube specimens (100X100X100 mm). The average of the experimental data from three identical specimens was utilized for all hardness testing results. Table 3 presents the hardened properties of concrete at 7 and 28 days, which were obtained for eight different mix proportions. For all the series, the compressive strength generally increased as the quantity of mineral admixtures raised.

Table 3. mechanical properties of concrete mixes

MIX	DENSITY	COMPRESSIVE STRENGTH in N/mm^2		UPV 28DAYS		Rebound number
		7 DAYS	28DAYS	VELOCITY in m/s	TIME in sec	28 days
DC0	2.47	39.04	46.43	4975	20.1	44
DC10	2.48	47.14	55.13	5181	19.3	48
DC20	2.49	56.17	61.00	5319	18.8	52
DC30	2.48	49.42	52.66	5236	19.1	46
DC40	2.46	39.83	44.91	5051	19.8	42
DC50	2.45	33.06	36.35	4926	20.3	38
DC60	2.45	25.61	28.54	4854	20.6	32
DC70	2.44	20.25	25.28	4695	21.3	30

5. Durability properties

Eight mix proportions performed the following durability tests: three cube specimens measuring 100 x 100 x 100 mm for each mix proportion were tested for water absorption after 28 days. Three samples of each mix percentage were made and examined after 28 days, with a sorptivity test performed on specimens measuring 100 mm in diameter and 50 mm in depth. We followed ASTM C-1202 for conducting the quick chloride ion permeability test. A 60-V potential was applied for six hours to three specimens that measured 100 mm in diameter and 50 mm in thickness and were prepared in accordance with the specifications. In order to assess each concrete mixture's chloride ion permeability, the total charge that travelled through the concrete specimens was calculated. The selection of hydrochloric acid (HCL) was based on concentration circumstances and the impact of an acidic environment on the characteristics of SCC. Every mix proportion was represented by two cube examples, each measuring 100 x 100 x 100 mm. 28-day cure period changes in compressive strength were examined.

5.1 Rapid chloride penetration test

Low-permeability concrete is required by requirements for many concrete structures constructed today. This test method is recognized by the construction industry as a means of assessing chloride permeability. After 28days of potable water curing, the quick chloride permeability test was used to evaluate the concrete's resistance to chloride penetration. Three specimens of each proportion, each one is placed between the two acrylic cells measuring 100 mm in diameter and 50 mm in height and having been conditioned in accordance with the standard were exposed to a 60 V potential for six hours. There are two chambers in the acrylic cell. One chamber contains sodium chloride solution, while the other chamber contains sodium hydroxide solution. 0.3M is the strength of sodium chloride solution, and 0.3M is the strength of sodium hydroxide solution. Because they establish a direct relationship between the volume of solution and the moles of solute, molar solutions are the most helpful for computing chemical reactions. The amount of salt is diluted to 1000 ml of solution in the following manner to create molar NaCl solutions of concentrations: One gram equals 1000 milliliters of water. Thirty grams of NaCl are dissolved in one liter of water to create a 3% NaCl solution. To create a 0.3M molar NaOH solution, a 1000 milliliter solution is used to dissolve 12 grams, of NaOH pellets. The diffuser cell has a power connecting terminal that is attached to a brass mesh to evenly distribute the current across the cross sectional area. The acrylic diffuser cell has a groove that allows the concrete module to be placed correctly. To stop chemical solution leaks, a brass washer and a neoprene rubber washer are inserted into the groove. Each of the four corners of the compartments has a fastening hole. Stainless steel bolts and nuts are used to secure the concrete specimen after it has been positioned between the two cells. Over the joints where the cell and the concrete meet, apply silicone grease. There can be a tiny space, which silicone grease was used to fill. Put on both ends. To stop chlorides from leaking through the surface of the concrete specimen, apply silicone grease to its whole surface. Give the silicone grease some time to dry. Following solidification, a 3 M strength sodium chloride solution is added to Cell1, the fitting black terminal, and it is meant to be connected to the RCPT apparatus's negative supply. 0.3M sodium hydroxide solution is placed in Cell 2 (which has a red terminal attached to it) and linked to the apparatus's 60 volt positive supply. There will be about 250 ml of fluid in each cell. To stop any leaks that may occur from gaps in the margins, silicon compound can be put to both the concrete specimen's surface and the edges. The cells are prepared to perform the chloride permeability test. If the computer is turned on, the logger enters the data. You can transfer the data to the computer by holding down the INC key for ten seconds straight. Every time a record is made, the data logger will capture the initial readings of the related cells. The system should stop after obtaining the last reading at the end of the log time. To assess the concrete's chloride permeability, the total charge that flowed through the specimens was calculated.

Calculation:

$Q = 900(I_0+2I_{10}+2I_{20}+2I_{30}+......+2I_{300}+2I_{330}+I_{360})$

Q = Charge passed (coulombs)

I_0 = Current passed through specimen immediately after voltage is passed

I_t = Current at t min after voltage is applied

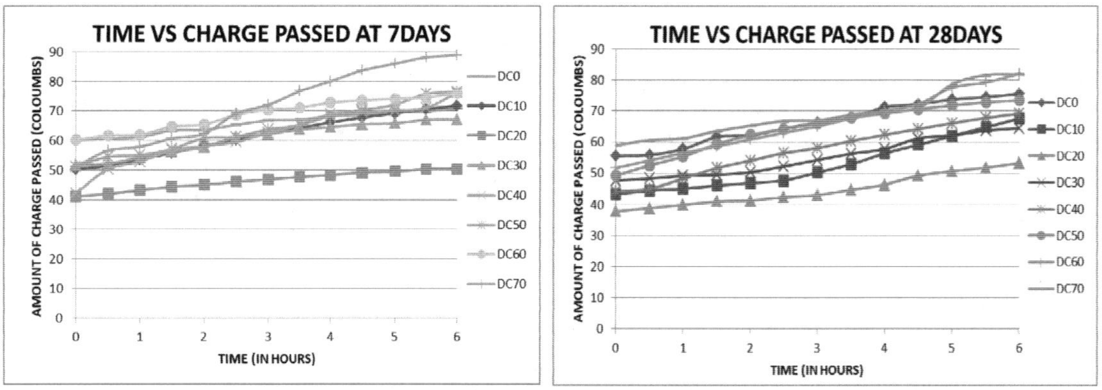

Figure 3. Average charge passed vs time graph at a) 7 days b) 28 days.

The test specimen's age and curing have a significant impact on the outcomes. If the sample has been thoroughly cured, the specimens have lower coulombs. Studies have also revealed that the inclusion of ionic salts in concrete admixtures may have an impact on the outcomes. The quantity of coulombs passed rises when ionic salts are present. Ionic salts are thought to provide an extra transport channel for the shift. Higher coulomb levels are the result, despite the fact that the permeability of the concrete remains unchanged. Ionic salt-containing additives are, in general, accelerators made of calcium nitrite, calcium nitrate, calcium chloride, and sodium thiocyanate. If this approach is to be used for testing concrete containing certain admixtures, tests must be conducted both with and without the admixture to determine the impact on the results. A partial list of additional variables that could impact the test results is as follows: Cement factor, air content, water cement ratio, test sample curing time, and kind or source of aggregate. According to ASTM C1202, "Standard test method for Electrical Indication of Concrete's Ability to Resist Chloride Ion Penetration," the chloride permeability of concrete is categorized as indicated in Table 4 based on the charge passed in coloumbs.

Table 4. Chloride penetration based on the charge passed

Charge Passed (Coulombs)	Chloride Permeability	Typical of
<100	Negligible	Polymer concrete, polymer impregnated concrete
100-1000	Very low	internally-sealed concrete or latex modified concrete
1000-2000	Low	conventional PCC have Low W/C ratio (<0.4)
2000-4000	Moderate	conventional PCC have Moderate W/C ratio (0.4-0.50)
>4000	High	conventional PCC have High W/C ratio(>0.60)

5.2 Sorptivity

According to ASTM 642-8, one of the main elements influencing the concrete's serviceability and durability is moisture infiltration through capillary action. Concrete specimens with a 100 mm diameter and 50 mm depth are cast in order to assess the sorptivity. After a 24-hour oven drying period and a 24-hour cooling period, the specimens' surfaces were sealed with silica gel, and their original weight was recorded. After submerging the specimens in water to a depth of 5 to 10 mm, the mass gain is measured at 30, 60, 90, and 120 minutes [2]. The sorptivity coefficient is computed by dividing the volume of water containing calcium hydroxide that enters the concrete per unit surface of contact by the square root of the exposure period of the prepared samples.

5.3 Water absorption

5.3.1 Initial water absorption
For this experiment, concrete specimens measuring 100 x 100 x 100 mm cube sizes were cast in accordance with ASTM C 642 requirements. The initial constant weight (W_1) of the concrete sample was obtained by oven drying them at 105°C. After being submerged in the curing water for 30 minutes at regular intervals (W_2), the specimens are weighed. Thirty minutes is the calculated starting water absorption.

5.3.2 Saturated absorption
For this experiment, concrete specimens measuring 100 x 100 x 100 mm cube sizes were cast in accordance with ASTM C 642 requirements. The initial constant weight W1 of the concrete specimens was obtained by oven drying them at 105°C. Following their immersion in the curing water, the specimens are weighed at regular intervals of 30 minutes (W_2) and 120 hours (W_3). At 120 hours, the saturation water absorption is computed [2].

5.4 Acid attack by HCL
Based on the concentration circumstances, the impact of an acidic environment on the characteristics of SCC using hydrochloric acid (HCL) was chosen. For the purpose of examining the characteristics of SCC subjected to a 5% solution of hydrochloric acid (HCL). For every mix percentage, two cubes were used. After 28 days of curing, the samples are taken out of the reservoir, dried with a cloth, and kept at room temperature for 24 hours. The mass and size of the samples are then documented, along with changes in their compressive strength, mass, and volume. In order to maintain absolutely equal test conditions and prevent salt buildup in the reservoir, the acid solution in this investigation was changed every ten days with fresh acid.

6. Results and discussion

6.1 Compressive strength and UPV test

The test findings at 7 days revealed that the proportions of the mix that contained mineral admixtures had a higher compressive strength than the proportion of the control mix, suggesting that the pozzolanic interactions between FA and DP were adequate to increase compressive strength at this early stage. However, the DP series showed the highest compressive strength results at 28 days due to the slower pozzolanic processes [4]. Compressive strength is increased in the cases of DP and FA by filling up the spaces between the bigger cement particles and enhancing the development of secondary hydrates through pozzolanic reactions with the lime left over from the primary hydration [17]. If the cement has a significant amount of tricalcium aluminate (C_3A), calcium carboaluminate will be produced from the reaction between calcium carbonate ($CaCO_3$) from the DP [15]. This reaction, accelerating the hydration and increasing the compressive strength, increases with the DP content of the cement. However, DP caused an increasing of early age performance of SCC. Ultrasonic pulse velocity has increases by increasing in compressive strength for all the mixtures. The reduction in the UPV was very nearly in line by way of the reduction in the compressive strength.

6.2 Rapid chloride penetration test

Figure 4 displays the total charge passing in 6 hours as a measure of the permeability of chloride. The mixtures utilized in this investigation were categorized using the ASTM C1202 recommended limits for chloride ion permeability. The chloride ion permeability of the SCC decreased as the replacement amount of dolomite mineral admixtures was increased. The charge that passes through the specimen increases when the proportion of dolomite rises over DC20, as can be seen from the data. However, there was a significant drop in charge at DC20. Thus, DC20 has the lowest charge going through specimens among the test findings shown in Figure 4. However, the pozzolanic reaction of fly ash causes an unexpected reduction in chloride ion penetration at 28 days. When the concentration of dolomite increases, the curing period increases when the penetration of chloride ions is slightly reduced. The findings indicated that the mixtures of DC60 and DC70 had a chloride ion permeability value of "low." Reduced chloride ion permeability depends on the development of a discontinuous pore system and a less porous, denser microstructure. It appears that the pozzolanic processes can form a discontinuous pore structure more easily [3]. Chlorides typically enter concrete through diffusion through porous pores or water pathways. Stable chlorocomplexes can be formed when certain chlorides combine with cement compounds, particularly tricalcium aluminates (C_3A). The corrosion process is started by the excess of free chlorides, which are free. Because there is more alumina in the mixture and more calcium silicate hydrate is produced during the pozzolanic reactions, the presence of FA and DP causes the amount of C_3A to increase. It is evident that the overall charge of the concrete is significantly impacted by the alumina concentration. The overall charge drops with increasing alumina (Al_2O_3) content [8]. The combinations that had higher concentrations of FA and DP, along with high alumina levels, demonstrated a significantly stronger resistance to the penetration of chloride ions.

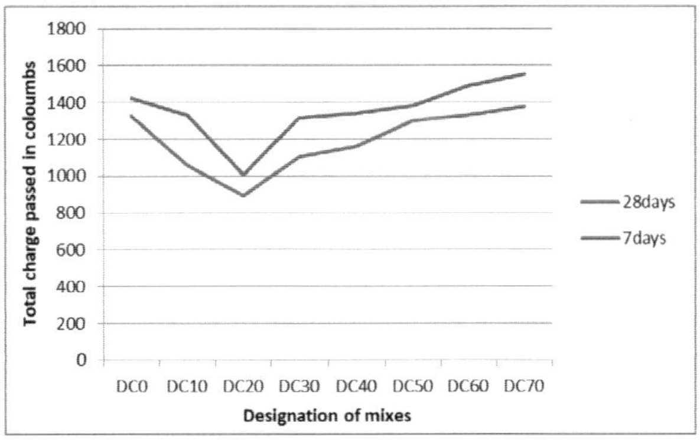

Figure 4. rapid chloride permeability of all mixes at 7 and 28days

6.3 Sorptivity

The sorptivity test results are displayed in Table 5, where it is abundantly clear that the microstructure was enhanced by the addition of DP and fly ash. Because of the homogenous nature of the SCC created in this study by the different sized elements, voids were mitigated, which reduced the water's capillary movement. When compared to control specimens, the denser medium absorbing rate in these specimens adequately justifies the applicability of the SCC thus created in moisture circumstances. When compared to the sorptivity coefficient of the control mix, the results showed that the sorptivity coefficient value was lower at the 20% replacement level.

Table 5. Sorptivity test results of all mixes.

Mix	Water penetrated weight through the contact surface		Volume of water penetrated/surface area of exposure		Sorptivity coefficient (k)	
	30min	120min	30min	120min	30 min	120min
DC0	4.25	9.15	0.054	0.116	0.010	0.011
DC10	3.35	7.65	0.043	0.097	0.008	0.009
DC20	2.65	7.05	0.034	0.090	0.006	0.008
DC30	3	8.2	0.038	0.104	0.007	0.010
DC40	4.65	9.6	0.059	0.122	0.011	0.011
DC50	6.1	10.3	0.078	0.131	0.014	0.012
DC60	6.75	12.5	0.086	0.159	0.016	0.015
DC70	7.48	14.64	0.095	0.186	0.017	0.017

6.4 water absorption

Table 6 summarizes the findings of the water absorption test. Next, a comparison is made between the results and the control sample. The water absorption percentage for DC20 is less than that of the control specimen. Additionally, according to the studies, there is a 20% replacement level reduction in initial water absorption with DP. The reason for this is that the finer fly ash and dolomite particles created a denser microstructure in the concrete specimens, which in turn decreased the number of voids and micro cracks in the concrete and produced a more durable concrete. The volume of voids and effective porosity has a strong relationship with the water absorption ratio.

Table 6. Initial water absorption, saturated water absorption and effective porosity test results of all mixes.

	initial water absorption at 30min	saturated water absorption at 120h	volume of voids	effective porosity
DC0	1.91	4.67	110	11
DC10	1.65	3.51	85	8.5
DC20	1.50	3.21	75	7.5
DC30	1.83	4.88	120	12
DC40	2.48	6.21	150	15
DC50	3.04	7.61	175	17.5
DC60	3.49	8.06	185	18.5
DC70	3.65	8.68	190	19

When the percentage of DP increases, the saturated water absorption falls until it reaches 20% and then rises to 70% replacement level. When it comes to DC20 mix, the volume of voids is minimal.

6.5 Acid attack by HCL

The compressive strength test results for all SCC specimens subjected to 5% HCL solutions for a maximum of 28 days are displayed in Table 7. It is discovered that, among various mixtures, the one with 70% DP substituted for cement results in the least drop in compressive strength. However, the specimen DC20 exhibits the highest compressive strength. Less than 40% of DP should be replaced with cement to show a positive effect; more than 40% should be replaced with cement to show a negative effect when it comes to HCL acid solution. When less than 20% of the mixture is incorporated by DP, it is demonstrated that the pozzolan reactions and hydration process are more comprehensive than when the combination contains more than 20% of DP. As a result, the process of making C-S-H gel is effective, results in concrete that is denser and more uniform, and minimizes pores. These elements lessen the samples' permeability and lessen the impact of acid attack on their strength [6]. The results show that, when placed in an acidic environment, SCC with a 20% drop in compressive strength had the least amount of reduction. After 28 days, the drop in compressive strengths of combinations with more than 20% (DC30, DC40, DC50, DC60, and DC70) relative to the control specimen falls between 10.39 and 16.33%.

Table 7. Initial water absorption, saturated water absorption and effective porosity test results of all mixes.

Mix	Before immersion weight	After immersion weight	Compressive strength after immersion in HCL	% of loss in strength	% of loss in mass
DC0	2.538	2.448	42.03	-9.487	-3.547
DC10	2.550	2.480	50.03	-9.260	-2.745
DC20	2.413	2.378	55.53	-8.975	-1.451
DC30	2.363	2.303	47.19	-10.392	-2.540
DC40	2.390	2.310	39.46	-12.130	-3.347
DC50	2.435	2.345	31.08	-14.498	-3.696
DC60	2.243	2.138	24.15	-15.391	-4.682
DC70	2.378	2.250	21.15	-16.337	-5.363

IOP Conf. Series: Materials Science and Engineering **1316** (2024) 012007 doi:10.1088/1757-899X/1316/1/012007

In table 7 Results for mass loss in samples exposed to HCL solution that contain DP and the control combination are displayed. The SCC sample results show that at a 70% cement replacement level with DP, the highest mass losses were attained. DC10, DC20, and DC30 combinations showed smaller mass loss and higher resistance in a 5% HCL solution compared to the control mixture. Furthermore, it was found that in the HCL acid solution, all combinations had a reduced mass loss [6]. There is no aggregate separation and just a small amount of the cement paste has broken down in the samples' top layer in the HCL solution. Based on visual observations, it was observed that the cube samples that were submerged in a 5% HCL solution had significantly reduced surface deterioration.

7. Conclusion

The following conclusions came about upon consideration of the experimental research work are

- The flow properties of concrete are achieved within the limits. The results of the slump flow is classified as SF2 as per the guidelines.
- Combination of dolomite powder and flyash improves the durability characteristics in addition to compressive strength.
- The performance of SCC mix containing dolomite powder as substitute to 20% cement is better than all other mixes including control mix. Study parameters such as rapid chloride penetration test, sorptivity, water absorption are improved in case of SCC mixes containg dolomite powder.
- Even at 40% replacement level of cement with dolomite powder, the compressive strength and durability parameters are almost equal to that of control mix.
- At 20% cement replacement level with dolomite powder, the SCC mix classified as "very low" as per ASTM C 1202-94 assessment criteria using rapid chloride penetration test. The range of 858 – 1377.3 coloumbs was reported for the chloride ion permeability test findings.
- The saturated water absorption of SCC mixes containing dolomite powder which is lower when compared with that of concrete mixes without having dolomite powder.
- The partial replacement of PC by FA and DP in SCC reduces the sorptivity coefficient upto 20% of dolomite powder. The sorptivity coefficient ranges from 0.008 to 0.017cm/min$^{0.5}$ were obtained from the test results.
- The percentage loss in compressive strength of SCC having dolomite, when subjected to acid attack test using HCL is within 10% upto the replacement level of 20%.

8. References

[1] Uysal M, Yilmaz K and Ipek M 2012 The effect of mineral admixtures on mechanical properties, chloride ion permeability and impermeability of self-compacting concrete *Construction and Building Materials* **27** 263–70

[2] Gnanaraj S C, Chokkalingam R B, Thankam G L and Pothinathan S K M 2021 Durability properties of self-compacting concrete developed with fly ash and ultra fine natural steatite powder *Journal of Materials Research and Technology* **13** 431–9

[3] Mohseni E, Miyandehi B M, Yang J and Yazdi M A 2015 Single and combined effects of nano-SiO2, nano-Al2O3 and nano-TiO2 on the mechanical, rheological and durability properties of self-compacting mortar containing fly ash *Construction and Building Materials* **84** 331–40

[4] Gupta N and Siddique R 2020 Durability characteristics of self-compacting concrete made with copper slag *Construction and Building Materials* **247** 118580

[5] Faraj R H, Sherwani A F H and Daraei A 2019 Mechanical, fracture and durability properties of self-compacting high strength concrete containing recycled polypropylene plastic particles *Journal of Building Engineering* **25** 100808

[6] Sharifi Y, Afshoon I, Nematollahzade M, Ghasemi M and Momeni M-A 2019 Effect of copper slag on the resistance characteristics of SCC exposed to the acidic environment *Asian Journal of Civil Engineering* **21** 597–609

[7] Frazão C, Camões A, Barros J and Gonçalves D 2015 Durability of steel fiber reinforced self-compacting concrete *Construction and Building Materials* **80** 155–66

[8] Wongkeo W, Thongsanitgarn P, Ngamjarurojana A and Chaipanich A 2014 Compressive strength and chloride resistance of self-compacting concrete containing high level fly ash and silica fume *Materials & Design* **64** 261–9

[9] Nguyen H-A, Chang T-P, Shih J-Y and Suryadi Djayaprabha H 2018 Enhancement of low-cement self-compacting concrete with dolomite powder *Construction and Building Materials* **161** 539–46

[10] Khan M I, Challa B, Varma S H and Abdy Sayyed M A 2019 Sorptivity and Durability Assessment of Dolomite Impregnated Ternary Concrete *International Journal of Recent Technology and Engineering (IJRTE)* **8** 5676–81

[11] Bakhtina T, Lyubomirskiy N and Bakhtin A 2019 Dolomite binding materials with low CO2 emission *Materials Today: Proceedings* **19** 1998–2004

[12] Shashikala A P, Nagarajan P and Parathi S 2019 MECHANICAL AND DURABILITY CHARACTERISTICS OF GGBS DOLOMITE GEOPOLYMER CONCRETE ed D Ozevin, H Ataei, M Modares, A P Gurgun, S Yazdani and A Singh *Proceedings of International Structural Engineering and Construction* **6**

[13] Poornima V, Ananthkumar M, Barathi Kumar S and Imthiyaz Ahamed K 2019 Suitability of dolomite silica-sand in partial replacement with m- sand for cement mortar *IOP Conference Series: Materials Science and Engineering* **577** 012135

[14] Saranya P, Nagarajan P and Shashikala A P 2021 Performance Studies on Steel Fiber–Reinforced GGBS-Dolomite Geopolymer Concrete *Journal of Materials in Civil Engineering* **33**

[15] Shaaban M 2021 Properties of concrete with binary binder system of calcined dolomite powder and rice husk ash *Heliyon* **7** e06311

[16] Barbhuiya S 2011 Effects of fly ash and dolomite powder on the properties of self-compacting concrete *Construction and Building Materials* **25** 3301–5

[17] Mikhailova O, Yakovlev G, Maeva I and Senkov S 2013 Effect of Dolomite Limestone Powder on the Compressive Strength of Concrete *Procedia Engineering* **57** 775–80.

[18] Yarramsetty N B B, Dudekula N R and Geddada N Y 2023 Experimental study on concrete strength properties by partial replacement of cement with Dolomite Powder & fine aggregate with M sand Materials today: proceedings.

IOP Conf. Series: Materials Science and Engineering **1316** (2024) 012008 doi:10.1088/1757-899X/1316/1/012008

A Study on γ-Ray Attenuation in MnO

A S Madhusudhan Rao[1], Parcha Kalyani[2], J Shankar[3]

[1, 3] Geethanjali College of Engineering and Technology, 501301, India
[2] Vignana Bharathi Institute of Technology, 501301, India

madhuammiraju@gmail.com

Abstract. An experimental study has been conducted to determine the attenuation of γ-photons in Manganese oxide (MnO). Determining the thermophysical properties of the chemical at various temperatures within the 300K–1075K temperature range is the primary goal of the study. Additionally, the investigation on attenuation has been expanded to assess photon interaction characteristics at varying γ-energies.

1. Introduction

High energy γ-photons' interactions with matter as a function of temperature are constantly fascinating and significant to many scientific disciplines.

Thermophysical parameters, such as density and thermal expansion, can characterize the physical state and behaviour of any material at any temperature. In addition to determining a material's thermal stability and thermal shock resistance, thermal expansion of solids also influences material selection for a variety of applications, including coolant materials, fuel materials for use in nuclear reactor safety studies, and post-accident heat transfer analysis of molten structural materials. Understanding the temperature fluctuation of many other properties, such as elastic constants, refractive indices, dielectric constants, thermal conductivity, diffusion coefficients, etc., depends heavily on the study of how these properties depend on temperature.

In order to ascertain the thermophysical properties of MnO, the attenuation analysis of γ-radiation at various temperatures makes use of the concept of Drotning [1]. Compared to other techniques such as X-ray diffraction, dilatrometric technique, computational approaches, etc., the current method has benefits [2- 3]. The study's determination of the mass attenuation coefficient (μ_m) allows for the investigation of the compound's thermophysical properties within the temperature range of 300K–1075K, as well as physical parameters.

Understanding physical factors such as the mass attenuation coefficient, atomic and electronic cross-section, among others, is beneficial when attempting to comprehend the characteristics of composite materials and compounds. In several practical domains, including radiation dosimetry, nuclear medicine, radiation protection, and nuclear diagnostics, their contribution is indispensable. The photon interaction parameters of diverse materials have been previously documented [4-7] by a variety of techniques. The author has utilized the narrow collimated beam transmission method [7] at varying γ-energies in this instance. Raju [8] previously conducted research on the temperature dependency of ultrasonic attenuation and related parameters for manganese oxide in the 300–500 K temperature range. The computations were examined for both shear and longitudinal waves propagating along various crystallographic axes. In addition to providing details regarding the internal organization, intrinsic, and anharmonic properties of the material, the elastic constants and density can be used to calculate ultrasonic velocities for longitudinal and shear waves. For this crystal, the thermal relaxation time and the non-linearity coupling parameters have also been determined. This work will provide a good an illustration of the ultrasonic attenuation behaviour in MnO and is helpful in characterizing the material. Grundy et al. [9] have evaluated the phase diagram for the system and the thermodynamic parameters of manganese oxides. Jacob et al. [10] determined the Gibbs energy of MnO production with accuracy (±0.25 kJ mol-1). Research has been done on the thermodynamics of the manganous-oxide system utilizing solid oxide electrolytes [11]. Using a dilatometric approach, the precise thermal expansion of manganosite (MnO) at temperatures ranging from 20°C to 850°C has been determined [12]. The manganosite was generated as a single crystal using a flame fusion process. The harmonic and anharmonic parameters of MnO are obtained by analyzing the collected data and refining Grüneisen's theory of thermal expansion.

1.1 Potential applications of MnO

Because of its wide range of applications, in the domains of science and technology, including high-density magnetic storage media, catalysts, ion exchange, molecular adsorption, electrochemical materials, varistors, and solar energy transformation, manganese oxide is essential to many different sectors. It is widely utilized in the manufacturing of ceramics, glass, lithium-ion batteries, poultry, fertilizers, textile printing, and food additives. Applications for manganese oxide in medicine and biology have attracted a lot of interest.

It has been observed from the literature that there aren't many reports on MnO at various temperatures. This inspired us to begin the investigation.

2. Experimental

Using 21 grams of powder and a hydraulic press operating at 2310 pounds per square inch, a sample pellet of MnO is created. The prepared pellet has a thickness of 1.35 cm. The author has gone into great depth about the experimental approach and setup specifics in [3, 13, and 14]. The method described in [7] is employed to determine the mass attenuation coefficient (μ_m) of the compound utilizing γ–photons from energy sources such as ^{137}Cs (0.66MeV), ^{60}Co (1.173MeV & 1.332MeV), and ^{241}Am (0.0595MeV). The radiation from the sources listed above is applied to the sample. Using detection equipment, the intensities transmitted with the sample (I) and without the sample (I_o) are recorded for ten minutes under the photo peaks of the Gaussian distribution [7]. Using γ–photons of ^{137}Cs (0.66MeV), the temperature dependency of thermophysical parameters such as density, linear attenuation coefficient, and thermal expansion of MnO has been studied, in accordance with the experimental protocol given [13]. γ-ray counts are detected and recorded at various temperatures within the range of 300K-1075K, both with and without sample.

3. Computational

Using Equations (1–9) of Section 3 [15], the mass attenuation coefficient and other physical characteristics for various photon energies are calculated. The thickness, inaccuracies in recorded intensities, and furnace temperature all contribute to the measured physical parameter uncertainty (Eqn. 10 [15]). Equations (1–10) of sections (2, 4) are used to calculate the linear attenuation coefficient, density (ρ), and coefficient of linear expansion (α) as functions of temperature [14].

4. Results and Discussion

Table 2 displays the mass attenuation coefficient (μ_m) values that were obtained by empirical relation [15] and compared with the values of X-Com [5] in the current investigation of MnO at various γ-energies. The mass attenuation coefficient is independent of the sample's physical state; it is also independent of temperature. Fig. 1 displays the values of (μ_m) from the current investigation. The likelihood of absorption decreases as incident energy rises since there is a greater chance of the photons passing through the sample.

Based on several factors such as counting statistics, efficiency errors, peak area evaluation, mass thickness measurements, experimental system, and others, the approximated error (Eqn. 19 [13]) in the experimental measurements is about 1.26%. In the range of 300K to 1075K, the temperature-dependent attenuation of ^{137}Cs (0.66MeV) γ-photons can be depicted using a quadratic polynomial, as shown in Figure 2.

$$(I_0/I)(T) = 1.39 + (2.19\times10^{-5})T + (-1.39\times10^{-7})T^2$$

IOP Conf. Series: Materials Science and Engineering **1316** (2024) 012008 doi:10.1088/1757-899X/1316/1/012008

Table 1 displays the values of MnO's linear attenuation coefficient, density, coefficients of linear and volume thermal expansions, and temperature range (300–1075 K). This is the first time that the values of μ_l and ρ as a function of temperature have been reported; no comparable data from the literature is available. It is evident from Table 2 and Fig. 3 that μ_l drop to 16.4 m^{-1} at 1075 K from 24.10 m^{-1} at 300 K. A polynomial of the second degree has been used to represent this variant

$$\mu_l(T) = (1.39) + (2.19\times10^{-5})T + (-1.393\times10^{-7})T^2$$

Since the linear attenuation coefficient is directly correlated with the material's density, changes in density with temperature are reflected in the coefficient.

Sample density decreases from 5180 kgm^{-3} at 300 K to 3522.80 kgm^{-3} at 1075 K (Fig.4), it can be expressed in the form of a quadratic equation

$$\rho(T) = (5199) + (4.41\times10^{-1})T + (-18.03\times10^{-4})T^2$$

Since the mass remains the same but the volume increases with the increase in temperature that results in decrease in density.

The rise in the equilibrium concentration of thermally produced Schottky faults is responsible for the density drop. Defects may arise from γ-ray irradiation. The coefficient of linear thermal expansion found in this study accords well with findings in the temperature range from other methods found in the literature (Fig. 4). The coefficient of linear thermal expansion values found in this investigation has been compared to those found using alternative methods [12].

This demonstrates that the fluctuation in density has not been impacted by γ-ray radiation. The equilibrium concentration of defects in MnO does not appear to have been impacted by the (30 mci) strength of the source or the (~5hrs) of irradiation in this work. Table -1 indicates that the thermal expansion (Δl/l) is proportional to temperature and may be expressed by a polynomial of degree two.

$$(\Delta L/L)(T) = -0.255 + (11.22\times10^{-3})T + (4.02\times10^{-6})T^2$$

The variation of coefficient of linear thermal expansion (α) can be defined by the equation given below

$$\alpha(T) = 9.723 + (6.65\times10^{-3})T + (-1.394\times10^{-6})T^2$$

Since the mean distance between molecules increases with temperature due to an increase in the amplitude of molecular vibrations, the coefficient of thermal expansion always rises with temperature. Additionally, there is a linear relationship between the CTE and the thermal capacitance per unit volume.

Physical parameters of MnO: linear attenuation coefficient (μ_l), total photon interaction cross-sections (σ_τ) and (σ_e), effective atomic number (Z_{eff}), effective electron number (N_{eff}) and photon mean free path (λ) are estimated by using the mass attenuation coefficients acquired in the current investigation at various γ-energies. Table 2 compares these values to the values estimated using semi-empirical relations (1–9) of section 3 [15] and X-Com. Z_{eff} and N_{eff} is shown to be independent of photon energy, staying constant. Table 2 indicates that the photon mean free path (λ) increases with photon energy. Utilizing the mass attenuation coefficients (μ_m) found in this study at various γ-energies, the physical parameters of MnO, such as the linear attenuation coefficient, total photon interaction cross-sections, effective atomic number, effective electron number, and photon mean free path, are estimated, using semi-empirical relations (1–9) Section 3 [15]. These values are shown in Table 2 and are found to be in good agreement. Z_{eff} and N_{eff} is shown to be independent of photon energy, staying constant. The photon mean free path (λ) increases as photon energy increases. This results from a decrease in the likelihood of photons interacting with the substance as energy increases.

Table 1- Attenuation, density and thermal expansion with temperature in MnO

T [K]	I_0/I	μ_l [m^{-1}]	ρ [kgm^{-3}]	(Δl/l) 10^{-3}	α [$10^{-6}K^{-1}$]	
					[PW]	Suzuki[12]
300	1.3846	24.1044	5180	3.441	11.47	11.5
325	1.38275	24.0053	5158.7	3.78988	11.6475	11.7
350	1.3809	23.9061	5137.39	4.13875	11.825	
375	1.37873	23.7896	5112.35	4.50438	12	12.1
400	1.37656	23.6731	5087.31	4.87	12.175	
425	1.37406	23.5383	5058.35	5.25425	12.3525	12.4
450	1.37156	23.4036	5029.4	5.6385	12.53	
475	1.36874	23.2512	4996.64	6.01175	12.65	12.7
500	1.36593	23.0988	4963.89	6.385	12.77	
525	1.36279	22.9282	4927.23	6.77025	12.89	12.9
550	1.35965	22.7576	4890.56	7.1555	13.01	
575	1.35618	22.568	4849.84	7.55725	13.1375	13.2
600	1.35271	22.3785	4809.11	7.959	13.265	
625	1.34892	22.1702	4764.35	8.35888	13.37	13.4
650	1.34512	21.9619	4719.58	8.75875	13.475	
675	1.34101	21.7345	4670.72	9.16563	13.575	13.6
700	1.33689	21.5072	4621.86	9.5725	13.675	
725	1.33245	21.2603	4568.82	9.98938	13.775	13.8
750	1.32801	21.0135	4515.77	10.4063	13.875	
775	1.32324	20.7466	4458.41	10.8371	13.98	14
800	1.31848	20.4797	4401.05	11.268	14.085	
825	1.31338	20.1926	4339.35	11.703	14.1825	14.2
850	1.30829	19.9054	4277.65	12.138	14.28	
875	1.30286	19.5967	4211.29	12.6053	14.4025	14.4
900	1.29743	19.2879	4144.94	13.0725	14.525	
925	1.29166	18.957	4073.83	13.552	14.6475	14.7
950	1.28589	18.6261	4002.72	14.0315	14.77	
975	1.2798	18.2735	3926.94	14.5033	14.8725	14.9
1000	1.27371	17.9209	3851.17	14.975	14.975	
1025	1.26728	17.5449	3770.38	15.4833	15.1025	15.2
1050	1.26085	17.169	3689.6	15.9915	15.23	
1075	1.2477	16.3929	3522.8	16.5228	15.37	15.4

Table 2- Comparison of Photon Interaction Parameters of MnO at different γ-energies

Am 0.0595MeV			Cs 0.66MeV			Co1.173MeV			Co 1.332MeV		
X-Com value	Empirical value	Expt. Value	X-Com value	Empirical value	Expt. Value	X-Com value	Empirical value	Expt. Value	X-Com value	Empirical value	Expt. Value
					μ_m [10^{-3} m^2kg^{-1}]						
88.12	88.1952	88.11	7.301	7.30153	7.299	5.507	5.50606	5.505	5.161	5.16162	5.16
					μ_l [m^{-1}]						
456.5	456.851	456.4	37.83	37.8219	37.81	28.53	28.5214	28.52	26.73	26.7372	26.73
					σ_t [10^{-25} barn/atom]						
51.89	51.9371	51.89	42.99	42.9979	42.98	32.43	32.4246	32.42	30.39	30.3962	30.39
					σ_e[10^{-26} barn/atom]						
30.71	30.7317	30.70	2.544	2.54423	2.543	1.919	1.9186	1.918	1.798	1.79857	1.798
					Z_{eff}						
16.9	16.9	16.9	16.9	16.9	16.9	16.9	16.9	16.9	16.9	16.9	16.9
					N_{eff}[10^{23} electron/g]						
2.870	2.870	2.870	2.870	2.870	2.870	2.870	2.870	2.870	2.870	2.870	2.870
					λ [10^{-2} m]						
0.219	0.21889	0.219	2.644	2.64397	2.645	3.506	3.50614	3.507	3.741	3.74011	3.741

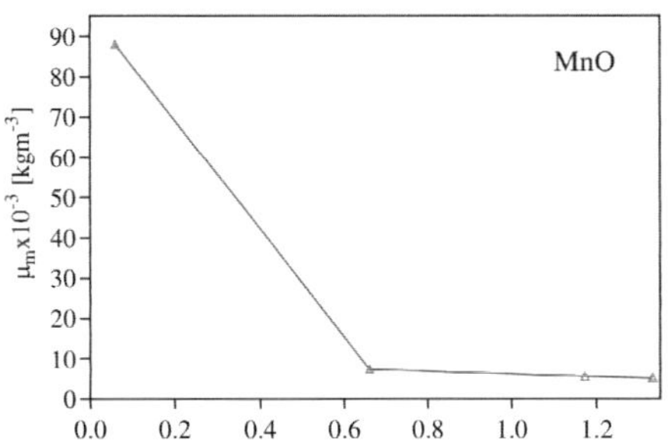

Figure 1. Mass attenuation coefficient (μ_m) VS Photon Energy.

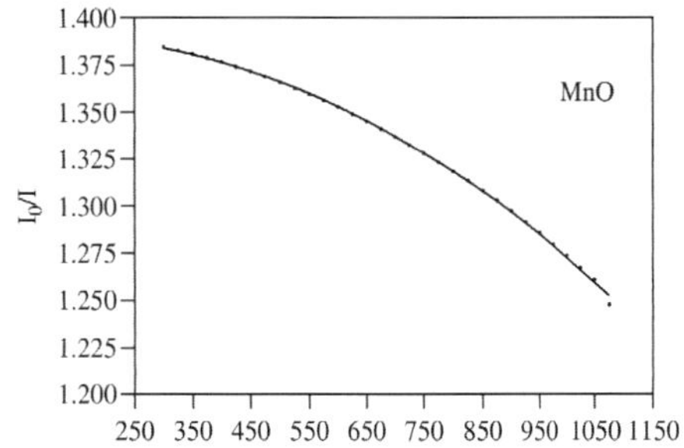

Figure 2. Temperature VS attenuation (I_0/I)

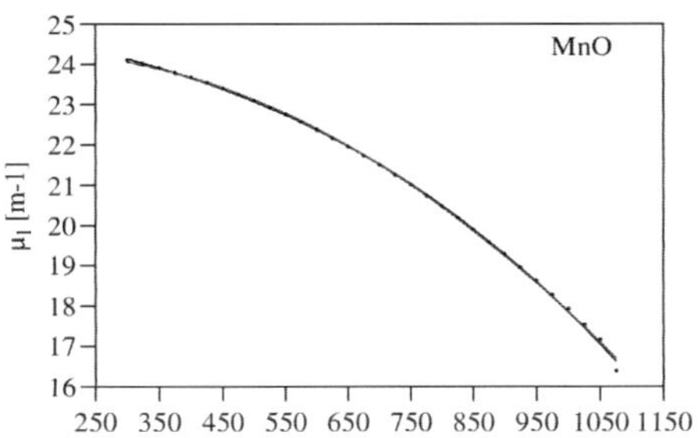

Figure 3. Temperature VS Linear attenuation coefficient (μ_l)

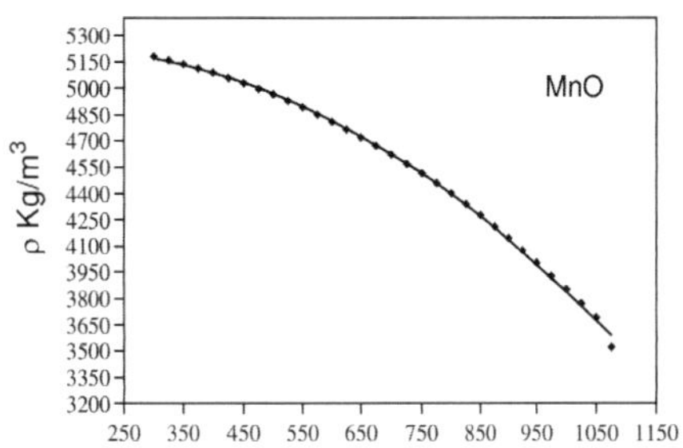

Figure 4. Temperature VS Density (ρ)

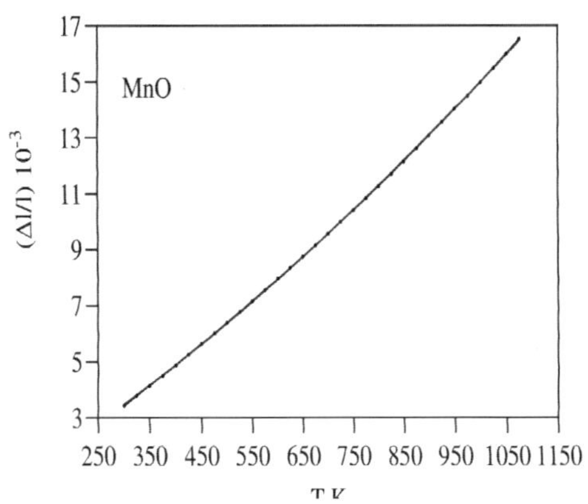

Figure 5. Temperature VS linear thermal expansion (ΔL/L)

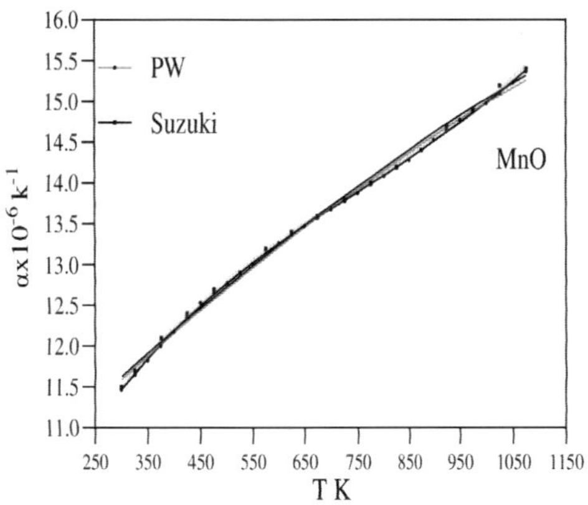

Figure 6. Comparison of Coefficient of linear thermal expansion (α) of present study with the data in the literature -Suzuki[12]

5. Conclusions

Using 21 grams of powder and a hydraulic press operating at 2310 pounds per square inch, a sample pellet of MnO is created. The prepared pellet has a thickness of 1.35 cm. Using a transmission beam approach, an experimental investigation of the attenuation of γ-photons at various energies has been carried out in MnO.For the first time, (μ_m) and additional parameters (σ_t, σ_e, Z_{eff}, N_{eff}, and λ) were estimated at various photon energies. We are aware that the (μ_m) is a sensitive and helpful physical measure that may be used to calculate a compound's (Z_{eff}) and (N_{eff}). As photon energy increases, the MnO values (μ_m) drop. The mass attenuation coefficient and (σ_t and σ_e) vary with energy in the same way.

The temperature dependence of mass attenuation coefficient, density, linear thermal expansion and the linear attenuation coefficient of the MnO were reported. Density and thermal expansion were shown to depend on temperature using second order polynomials. The results of this work's calculation of the coefficient of thermal expansion are in good agreement with values reported in earlier research published in the literature. For the first time, MnO has been subjected to the γ-ray attenuation technique. This is the first time that the values of density and linear attenuation coefficient as a function of temperature have been reported; no comparable data from the literature is available.

References

[1] DrotningW. D,"Thermal Expansion of Glasses in the Solid and Liquid Phases", International J of Thermophysics,Vol. 6, 1985.

[2] DrotningWilliam.D, "Thermal expansion of solids at high temperatures by the gamma attenuation technique", Rev Sci Instrum, 50, No12, 121, 567. 1979.

[3] MadhusudhanRao A. S et.al. "Thermo physical Properties of NaCl, NaBr, NaF by Gamma Ray Attenuation Technique", J Modn Phys, Vol.4, No.2, 208-214., 2013.

[4] Hubbell, J.H., Seltzer, S.M., Int. J. Appl.Radiat. Isot. 33, 1269–1290, 1995.

[5] Gerward L, Guilbert N, Bjorn Jensen K and Levring H, Radiat. Phys.Chem. 60, 23-24, 2001.

[6] Preseren R and Korde A, Radiation Physics and Chemistry, 55, 363, 1999.

[7] MadhusudhanRao A. S, et al "Study of Energy Dependent γ- ray Attenuation and Determination of Effective Atomic Number and Electron Density of Sodium Halides", Journal of Applied Physical Science International. Vol. 7 (4). 199-206, March, 2017.

[8] Krishna Murti Raju," Behavior of Ultrasonic Attenuation in MnO", Open Journal of Acoustics, 3, 54-59, September, 2013, http://dx.doi.org/10.4236/oja.2013.33A009.

[9] Grundy A.N, Hallstedt B and Gauckler, L.J "Assessment of the Mn-O system"J. Phase Equilib., 24, 7-39, 2003.

[10] Jacob K .T, Kumar A and Waseda Y, "Gibbs energy of formation of MnO" Journal of Phase Equilibria and Diffusion. 29, 222-230, 2008.

[11] Alcock C. B and Zador S," Thermodynamic study of the manganese/ manganous-oxide system by the use of solid oxide electrolytes" Electrochimica. Acta, 12, 673-677, 1967.

[12] Isao SuzuKI, Shin-ichi OKAJiua, and Kiyoshi SEYA, "Thermal expansion of single crystal Manganosite", J. Phys. Earth, 27, 63-69, 1979. https://doi.org/10.4294/jpe1952.27.63.

[13] MadhusudhanRao A. S et.al. "Thermo physical Properties of Rubidium and Lithium Halides by Gamma Ray attenuation Technique", Journal of High Temperature, Springer Link, Issue-5, 2014.

[14] MadhusudhanRao A. S et.al "Studies on Thermo physical Properties of CaO and MgO by Gamma Ray attenuation", Journal of Thermodynamics, April, 2014.

[15] MadhusudhanRao A.S et al, "Determination of Photon Interaction Parameters of CaO and MgO for Multi-Energetic Photons using γ-Ray Attenuation Technique", IOSR-JAP, Volume 8, Issue 2 Ver. I ,103-109, Mar. - Apr. 2016.

Compressive strength study of fibre reinforced concrete

Y V Bhavani Sai Kumar[1,2], R Padmapriya[3]

[1]Sathyabama Institute of Science and Technology (Deemed to be University), Chennai, Tamil Nādu - 600119, India
[2]Sasi Institute of Technology and Engineering, Tadepalligudem Andhra Pradesh - 534101 India
[3]School of Building and Environment, Sathyabama Institute of Science and Technology (Deemed to be University) Chennai, Tamil Nādu - 600119 India
padmapriya.civil@sathyabama.ac.in

Abstract: Materials play a vital role in the development of human life. Environmental and sustainable development concerns have seriously affected the development and applications of materials. Due to the advancement of rapid growth, it leads to the utilization of more materials, which leads to the depletion of natural materials. So, this leads the researchers to focus on the materials that have less impact on the environment. This study focuses on the compressive strength-based study of M30-grade fiber-reinforced concrete, partially replacing cement with flyash, fine aggregate with cocopeat, and adding coir fiber as a fiber. In recent times, there has been a tremendous increase in the usage of several natural fibers because of their low cost, ease of processing, and environmental friendliness. Due to their availability and eco friendliness, they are used in the development of fiber-reinforced concrete to improve its strength. The experiment entails casting and curing of different concrete mixtures for 3, 7, and 28 days with varying percentages of coconut fiber and cocopeat, 30% replacement of cement with flyash, and performing compressive strength tests in accordance with IS standards. Test results show adding 1% of coir fiber attains more strength in comparison with normal concrete and concrete with partial replacement of cement with fly ash and fine aggregate with cocopeat.

Keywords: *Compressive strength, fibre reinforced concrete, cocopeat, flyash, coir fiber*

1.Introduction:

Over the past few years, there has been a tremendous increase in the negative impacts on the environment, primarily due to the release of greenhouse gas emissions. Carbon dioxide is one of the major greenhouse gases that have negative impacts on the environment [1]. Among all the industries, the construction sector is one of the major contributors to the release of greenhouse gases. To reduce the negative impacts on the environment due to the release of greenhouse gases, researchers are focusing on net zero-carbon emissions. [1]. This made the researchers focus on eco-friendly construction materials [2]. Materials from agricultural and forest wastes, construction wastes, and demolition wastes have grabbed the attention of researchers [1].

Concrete is one of the most important materials used in construction activities when compared to other materials because of its various feasible properties, availability of raw materials, and low cost. The raw materials for the manufacturing of concrete are cement, fine aggregate, and coarse aggregate [3]. Apart from its advantages, concrete production is one of the contributors to greenhouse gas emissions, and it also consumes more natural resources. [4]. To reduce the negative impacts on the environment by construction materials, researchers are focusing on alternative raw materials, waste recycling, and the utilization of natural fibers in construction materials. Application of natural fibers in construction industry way back 3000 years ago with the use of straw in clay and the research on natural fiber-reinforced concrete way back from the late 1960's [5]. Natural fibers are classified into plant- and animal-based fibers, and plant-based fibers are composed of cellulose. In recent times, a lot of research has been conducted on the usage of plant-based fibers in the load-bearing components of construction [6]. Plant fibers include coconut fibers, sisal, jute, carp, and cannabis; eucalyptus grandis; malva; ramie past; pineapple leaves; cannabis; aspheric leaves; abaca leaves; jaws; dates; bamboo; banana; palm; hemp; linen; cotton; and sugar cane [7].

Normal concrete is low-tensile-strength and brittle in nature [8]. Steel embedded in plain concrete and the use of natural and synthetic fibers can enhance the properties of plane concrete [5]. When short, randomly distributed fibers are used to reinforce the matrix (paste, mortar, and concrete), a gradual increase in the tensile strength is observed [8]. The use of natural fibers in concrete is becoming more popular due to its low cost and having less impact on the environment [5]. Natural fiber-reinforced concrete has high strength and ductility when compared to plain concrete [5]. A tremendous amount of research is done by the various researchers on the coconut fiber because of its strength and its ability to take strains when compared to the other fibers [2, 5, 7]. Adverse effects on the environment can be reduced during concrete production by reducing the usage of cement, which emits more carbon content, by partially replacing cement with flyash without compromising its durability aspects [9]. The consumption of natural resources for the production of concrete can also be reduced by partially replacing the raw materials for the production of concrete with alternative materials. one such material usage of unutilized cocopeat in concrete production as a partial replacement of sand, since disposal of agricultural waste is creating a lot of problems and showing negative effects on the environment [3].

Therefore, this research work mainly focusses on the study of compressive strength characteristics of M30 grade fibre reinforced concrete partially replacing cement with flyash and fine aggregate with cocopeat. The fibre used in this research work is coconut fibre.

2. Materials and Methods:

The following are the raw materials used for the production of M30 grade of concrete: Cement, Flyash, Coconut fibres, Cocopeat, fine aggregates(sand), Coarse aggregates, super plasticizer and the raw materials are procured form the local vendors.

2.1 Cement:

Ordinary Portland cement commonly used raw material for the production of concrete across the world [1], it is a controlled mixture of calcium silicates, aluminates and ferrate which is ground to a fine powder with gypsum and other materials [11]. Therefore, the present uses OPC 53 grade of cement which is having a specific gravity of 3.10.

2.2 Fine aggregate:

The sand passing through 4.75mm sieve and retained on 600 µm sieve comes under the zone -II and as per IS:383-970 was used and it was free from sand silt and organic matter. The sand which was used is having a specific gravity of 2.67.

2.3 Coarse aggregates:

Coarse aggregate is the important raw material of the concrete as it forms the body of the concrete [10].20mm size aggregates procured form the local vendors were used throughout the research work having specific gravity of 2.50 as shown in the 'figure 1'.

'Figure 1' Coarse aggregate used in the study

2.4 Flyash:

The use of flyash is becoming important to engineers because of its economic and environmental benefits, and its usage as a partial replacement for cement can be limited based on the amount of free lime in the ash [9]. Several research investigations have been conducted to investigate the use of different percentages of flyash in concrete mixes as a partial replacement of cement, ranging from 5% to 60% [9, 12]. Flyash used in this work was procured from the local vendor with a 30% partial replacement of flyash in cement.

2.5 Cocopeat:

India is one of the largest producers of coconuts across the world, with a production of 19,309.90 million nuts during the years 2021–22 [13]. Coconut consists of 50% of the husk, which is used in the fiber production process, during which it extracts husk pieces, fiber, and peat. From this, only 11% of the husk is useful, and the remaining husk and peat pieces are removed as waste, causing serious environmental problems. Cocopeat, which is produced during the extraction of coconut, has very limited applications; the majority of it is disposed of [14]. Its production is about 7.5 million tons annually in India, and it is dark brown in color with particle sizes ranging from 0.2-4 mm [3]. In this present work, cocopeat is used as a partial replacement for fine aggregate in the production of M30-grade concrete which is as shown in the 'figure 2'. Cocopeat is treated with an alkaline solution to improve its durability in cement paste [3, 15].

'Figure 2' Cocopeat used in the work

2.6 Coir fibres:
Coir fiber is obtained from the husk of coconut, which occupies a predominate position in the production of the crop with an annual yield of about 50–100 fruits per tree. There is a gradual growth in the production of coir fiber due to increasing awareness of sustainable and biodegradable natural materials [16]. Coir fiber is composed of cellulose, cellulose sugar, and lignin as its main composition [17, 7, 18]. It is treated with chemicals to improve its properties and bond with the constituents [16, 17, 18]. In this work, untreated coir fibers procured from local vendors were used.

Concrete specimens were cast with partially replacing cement with flyash of about 30% and fine aggregate with cocopeat in the proportions of 5%, 10%, 15%, and 20%. In addition, coir fibers are added to the concrete mix in proportions of (1% and 1.5%) to the weight of the cement and tested for compressive strength for 3, 7, and 28 days.

3.Experimental work:
The raw materials are procured and then made ready for the production of M30-grade concrete, and the mix proportions were obtained as per the Indian standard specifications [19]. First, the cocopeat underwent an alkaline treatment by being submerged in a NaOH solution, cleaned with water, and then dried [3, 15], followed by the preparation of cement, fine aggregates, coconut fibers, and superplasticizer in appropriate quantities for the production of M30-grade concrete. Concrete specimens were made as per the mix proportions by mixing, casting, and curing for 3, 7, and 28 days, and tested for compressive strength, which is one of the mechanical properties of concrete. A comparative study in terms of compressive strength is made between the normal M30 grade and the concrete specimens prepared by partially replacing cement with flyash, fine aggregate with cocopeat, and adding coir fiber.

4.Results and Discussions:
A comparative study on the compressive strength was made on the coconut fiber-reinforced concrete M30 grade with and without partial replacements of cement with flyash and fine aggregate with cocopeat and the concrete specimens are shown in the 'figure 3'. For a normal M30 grade of concrete, a compressive strength of 38.21 N/mm^2 was observed for 28 days, and with the addition of coconut fiber, it increased by 1%. The strength was increased to 40.74 N/mm$^{2.}$ An increase in early strength was observed when 30% of cement was partially replaced with flyash, and the 28-day strength was 39.25 N/mm2, which is less than the normal coconut fiber-reinforced concrete with 1% coconut fiber addition. Concrete with 30% cement replacement with fly ash and addition of fibers by 1 and 1.5% shows almost similar values of compressive strength for 28 days as of normal M30 grade concrete. Replacement of cocopeat partially by 10, 15, and 20% to fine aggregate shows a decrease in compressive strength values

for 28 days but attains an almost similar value as of normal M30 grade mix concrete for 5% replacement. The results were tabulated below table 1 and shown in 'figure 4'.

Table-1: Compressive strength values of concrete specimens

S.No	Day	Mix	Strength of the Specimen(N/mm^2)
1	3	Normal M30 grade mix	17.32
	7		25.77
	28		38.21
2	3	Normal M30 grade with 1% coconut fiber	15.7
	7		26.6
	28		40.74
3	3	30% flyash partial replacement of cement	18.51
	7		27.4
	28		39.25
4	3	30% flyash partial replacement of cement and 1% coconut fiber addition	16.14
	7		22.66
	28		38.95
5	3	30% flyash partial replacement of cement and 1.5% coconut fiber addition	16.59
	7		22.96
	28		38.06
6	3	30% flyash partial replacement of cement and 1% coconut fiber addition and 20% cocopeat partial replacement of fine aggregate	2.36
	7		8.44
	28		23.18
7	3	30% flyash partial replacement of cement and 1% coconut fiber addition and 15% cocopeat partial replacement of fine aggregate	3.84
	7		11.55
	28		26.36
8	3	30% flyash partial replacement of cement and 1% coconut fiber addition and 10% cocopeat partial replacement of fine aggregate	9.03
	7		18.81
	28		30.07
9	3	30% flyash partial replacement of cement and 1% coconut fiber addition and 5% cocopeat partial replacement of fine aggregate	15.25
	7		24.88
	28		38.36

IOP Conf. Series: Materials Science and Engineering **1316** (2024) 012009 doi:10.1088/1757-899X/1316/1/012009

'Figure 3' Concrete Specimens

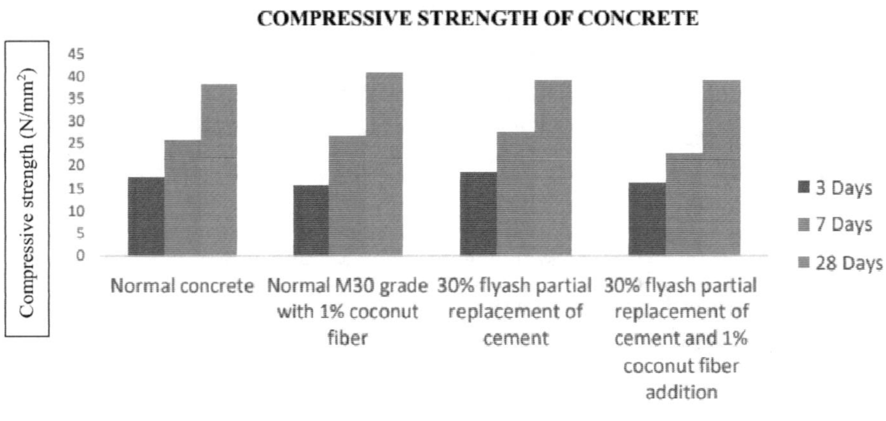

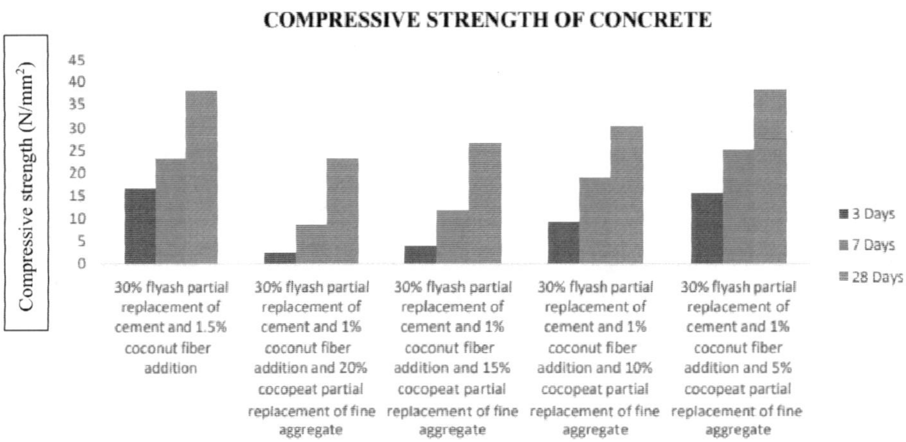

'Figure 4' Compressive strength of concrete without and without replacement of material

5. Conclusions:

The compressive strength of coconut fiber-reinforced concrete specimens incorporating cocopeat as a partial replacement of fine aggregate and flyash as a partial replacement of flyash for trial mix proportions was determined. The following conclusions are drawn:

1. Using cocopeat as a partial replacement for fine aggregate (sand) can reduce the cost of construction and reduce the usage of non-renewable resources like sand.
2. Using flyash as a partial replacement for cement can reduce the utilization of cement in concrete production, and the cost of construction can be reduced.
3. Using the partial replacement of fly ash and cocopeat with cement and fine aggregate in concrete production makes the concrete more conducive to the development of sustainable materials because it reduces the utilization of cement in concrete production, as cement production is one of the major contributors to the release of greenhouse gases and can reduce the depletion of non-renewable source sand.
4. The optimum percentage of replacement of fine aggregate by cocopeat is obtained at 5%. The maximum compressive strength at the optimum percentage is obtained at 38.36 N/mm2.
5. The optimum percentage of coconut fiber is obtained at 1%. The maximum compressive strength at the optimum percentage is obtained at 40.74 N/mm2.
6. The compressive strength of the coconut fiber-reinforced concrete by partially replacing cement with flyash and fine aggregate with cocopeat is greater than that of normal concrete up to an optimal replacement of 30% flyash, 5% cocopeat, and 1% coconut fiber.

References:

[1] Wang B, Yan L, Kasal B. A review of coir fibre and coir fibre reinforced cement-based composite materials (2000–2021). *J Clean Prod.* 2022 Mar 1;**338**:130676

[2] Ardanuy M, Claramunt J, Toledo Filho RD. Cellulosic fiber reinforced cement-based composites: A review of recent research. *Constr Build Mater.* 2015 Mar15;**79**:115- 28.

[3] Priyadarshini V, Felixkala T, Depaa RA, Hemamalinie A, Xavier JF, Babu KS, Arthi AJ.Experimental investigation on properties of coir pith and its influence as partial replacement of fine aggregate in concrete. *Mater Today-proc.* 2021 Jan 1;**45**:6903-6.

[4] Adesina A, Zhang J. Impact of concrete structures durability on its sustainability and climate resiliency. Next Sustainability. 2024 Mar 28:100025.

[5] Wang W, Zhang Y, Mo Z, Chouw N, Jayaraman K, Xu ZD. A critical review on the properties of natural fibre reinforced concrete composites subjected to impact loading. *Journal of Building Engineering.* 2023 Aug 10:107497.

[6] John MJ, Thomas S. Biofibres and biocomposites. *Carbohyd Polym.* 2008 Feb 8;**71**(3):343-64.

[7] Syed H, Nerella R, Madduru SR. Role of coconut coir fiber in concrete. *Mater Today-Proc.* 2020 Jan 1;**27**:1104-10.

[8] Revathi S, Dinesh M, Varsan SS. Mechanical properties of concrete incorporating coconut fibers and copper slag. *Mater Today-Proc.* 2023 Mar 2.

[9] Nayak DK, Abhilash PP, Singh R, Kumar R, Kumar V. Fly ash for sustainable construction: A review of fly ash concrete and its beneficial use case studies. *J Clean Prod.* 2022

Dec1;**6**:100143.

[10] V Shenbaga kumar, K Udhaya kumar, R Ramesh, R Ram Kumar, Mrs. N. Venkateshwari an experimental study on behaviour of concrete with coco-peat. *IRJET*, Volume: **06** Issue: 03 | Mar 2019.

[11] M.S. Krishna Hygrive, I. Siva Kishore, KJB Chari comparative study on compressive strength of flyash concrete, *IJCIET* Volume **8**, Issue 4, April 2017.

[12] Krithika J, Kumar GR. Influence of fly ash on concrete–A systematic review. *Mater Today-Proc*. 2020 Jan 1;**33**:906-11.

[13] Coconut Development Board [CDB 21-22] [CDB - Statistics (coconutboard.gov.in)] last visited on 04/04/2024.

[14] Sathiparan N, Anburuvel A, Muralitharan M, Kothalawala DA. Sustainable use of coco pith in cement-sand mortar for masonry block production: Mechanical characteristics, durability and environmental benefit. *J Clean Prod*. 2022 Aug 1;**360**:132243.

[15] Corradini E, de Morais LC, de F. Rosa M, Mazzetto SE, Mattoso LH, Agnelli JA. A preliminary study for the use of natural fibers as reinforcement in starch-gluten-glycerol matrix. In Macromolecular symposia 2006 Dec (Vol. **245**, No. 1, pp. 558-564).Weinheim: Wiley-Vch Verlag.

[16] Mishra L, Basu G. Coconut fibre: its structure, properties and applications. In Handbook of natural fibres 2020 Jan **1** (pp. 231-255). *Woodh Pub*

[17] Yew BS, Muhamad M, Mohamed SB, Wee FH. Effect of alkaline treatment on structural characterisation, thermal degradation and water absorption ability of coir fibre polymer composites. Sains Malays. 2019 Mar 1;**48**(3):653-9.

[18] Koňáková D, Vejmelkova E, Čáchová M, Siddique JA, Polozhiy K, Reiterman P, Keppert M, Černý R. Treated coconut coir pith as component of cementitious materials. Advances in *Mater SCI ENG*. 2015 Oct 1;2015.

[19] IS:10262-2019. Indian Standard Concrete Mix Proportioning – Guidelines (Second Revision). Bureau of Indian Standards, New Delhi, India.

IOP Conf. Series: Materials Science and Engineering **1316** (2024) 012010 doi:10.1088/1757-899X/1316/1/012010

Study of structural, dielectric properties of $ZnFe_2O_4$: effect of thermal annealing

S S Modak[1], C Parmar[1], R Verma[1], V R Reddy[2] and S N Kane[1,*]

[1] Magnetic Materials Laboratory, School of Physics, D. A. University, Khandwa Road, Indore-452001, India
[2] UGC-DAE Consortium for Scientific Research, Khandwa Road, Indore-452001, India

[*]Corresponding Author Email: kane_sn@yahoo.com

Abstract. We report structural, dielectric properties of sol-gel auto-combustion-synthesized $ZnFe_2O_4$ nano-ferrites (grain-size: 53.43 - 64.81 nm), and study the effect of thermal annealing on structural dielectric properties by X-ray diffraction (XRD), dielectric measurements. Results show that thermal-annealing shows: i) an increase of lattice parameter ($a_{exp.}$), and on B-site: there is an increase of Zn^{2+} population with concurrent decrease of Fe^{3+} ions, ii) alterations of inversion-degree from 0.18 to 0.32, strengthening of A-O-B, A-O-A interactions, accompanied by a weakening of B-O-B super-exchange-interaction, iii) decrease of theoretical magnetization at 0 K ($M_{s(th)}$), iv) with increasing frequency, real-imaginary components (ε', ε'') of dielectric constant, and loss-tangent decreases, v) that at higher frequencies, ac conductivity (σ_{ac}) shows a significant increase and exhibited a power law dependence, suggesting a typical of charge transport assisted by a hopping or tunneling processes. Results also reveal that polarizability strength decrease with annealing temperature, and correlated-barrier-hopping remains the dominant mechanism for conduction. Cole-Cole plots (ε',ε''), corresponding complex impedance (Z',Z'') data are indicative of the growing influence of grain boundaries on the dielectric properties for the studied samples. Observed improved-response of σ_{ac}, recommends the potential-applications of the studied samples in high-frequency device applications.

1. Introduction

Spinel ferrites, denoted by the general formula $MeFe_2O_4$, where 'Me' represents a divalent metal ion are classified under the Fd_3m space group. They exhibit a face-centered cubic structure, characterized by two interpenetrating sub-lattices: A, situated at tetrahedral sites, and B, located at octahedral sites [1]. The structural, magnetic, and electronic properties of ferrites can be finely adjusted by manipulating synthesis conditions [2], modifying cationic distribution through compositional changes [3], incorporating specific metal ions such as Zn, Ni, Mg, or Co, among others [4-9], and employing post-preparation thermal, and irradiation treatments [3, 5-8]. It has been observed that addition of Zn plays a decisive role in tuning the functional properties by affecting the cationic distribution. As expected otherwise replacement of magnetic ion by a non-magnetic ion, increases saturation magnetization [1] as the non-magnetic ion Zn^{2+}, preferentially occupies A site, forcing Fe^{3+}ions to migrate to B site thereby increasing the net magnetic moment. Moreover, the presence of Zn has observed to improve the electromagnetic properties, such as initial permeability, saturation magnetization, and impedance [10, 11]. The incorporation of Zn content notably enhances density, magnetic moment, and average grain size [12, 13].

It is worth noting that in this study, we have endeavoured to calculate the complex impedance spectra based on the complex permittivity data. Complex impedance analysis is a valuable tool for examining electrical properties, offering insights into various regions such as grain, grain boundaries, and electrode interfaces in nanocrystalline systems across a range of applied frequencies. To the author's understanding, there are no significant reports available on the study of ac conductivity mechanism using calculated impedance spectra.

Thus, objective of this work is to explore the structural changes induced by thermal treatment, and their influence on the dielectric behaviour of Zn-ferrite, prepared using the sol-gel auto-combustion method. This investigation also aims to understand the impact of Zn on both the structure, and dielectric behaviour of the examined samples, thereby enhancing our understanding of the conductivity mechanisms for potential optimization in various frequency-dependent applications.

2. Experimental Details

Zn-ferrite ($ZnFe_2O_4$) nanoparticles were synthesized by sol-gel-auto combustion method, by taking AR grade nitrate-citrate precursors [Zinc Nitrate – $Zn(NO_3)_2{\cdot}6H_2O$, Ferric Nitrate –$Fe(NO_3)_3.9H_2O$, and Citric Acid: $C_6H_8O_7$]. All the precursors were taken in stoichiometric ratio, and were mixed in 10 ml de-ionized water. Then pH 7 was maintained by adding ammonia [NH_4OH] solution, and keeping metal salt to fuel (citric acid: initially acts as chelating agent and then as fuel) ratio was kept as 1:1. Obtained solution was heated at ~ 100 °C in air until fluffy powder was formed, called as 'as-burnt' or 'dry gel' powder. Prepared as-burnt powder was thermally treated in air at 400 °C, 500 °C, and 600 °C for 2 hours respectively, and was used for structural, and dielectric characterization. The X-ray diffraction (XRD) data were obtained using a Bruker D2 Phaser advanced diffractometer, which employs Cu-Kα Radiation with a wavelength of 0.15418 nm. The measurements conducted in a θ-2θ configuration at room temperature.

XRD patterns were analysed to calculate structural parameters: including the experimental lattice parameter '$a_{exp.}$' (Determined using the (311) reflection) and the x-ray density (ρ_{xrd}) as referenced in [14]. The determination of grain diameter ($D_{W\text{-}H}$), and strain (ε) was achieved by incorporating both instrumental broadening (obtained from XRD data of a standard LaB_6 sample), and strain broadening, utilizing the Williamson Hall (W-H) method [15, 16]. Additionally, the dislocation density (ρ_D), and specific surface area (S) were computed based on the grain diameter ($D_{W\text{-}H}$) values, as outlined in [17]. Dielectric properties (ε′,ε″), loss tangent (tan δ), and ac conductivity (σ_{ac}) of the samples in parallel plate capacitor geometry with the ferrite samples as the dielectric medium (10 mm diameter pellets) in the frequency range: 100 Hz – 1 MHz, were measured using a LCR meter (Agilent Technologies: model E4980A).

Cationic distribution was determined using the Bertaut method [18], which involves comparing experimental, and calculated/computed intensity ratios for (200), (400), and (422) planes, as outlined in [19]. This method provides insight into the distribution of cations on A, and B sites. The intensity ratios, such as $I_{(220)}/I_{(440)}$, $I_{(400)}/I_{(422)}$, and $I_{(220)}/I_{(400)}$, vary depending on the cation distribution on these sites. The optimal distribution of cations on the A, and B sites is determined by comparing the theoretical, and experimental ratios (I_{hkl}^{Calc} and I_{hkl}^{Obs}) of the calculated, and observed intensities, respectively, ensuring noticeable agreement. The derived cationic distribution was utilized to calculate the theoretical magnetization ($M_{s(th)}$) at 0 K, also referred to as the Néel magnetic moment (n_N) [17]. Additionally, oxygen position parameter (u), and bond angles (θ_1, θ_2, θ_3, θ_4, θ_5) were also determined using the cationic distribution data [16].

Following expressions were utilized to determine the frequency-dependent dielectric permittivity (ε', ε"), loss tangent (tan δ), and AC conductivity (σ_{ac}) [20, 21]:

$$\varepsilon' = C_P \times d/\varepsilon_0 \times A \quad (1)$$

where C_p – Parallel Capacitance, d – pellet thickness A – pellet cross-sectional area, ε_o– permittivity of free space.

$$\varepsilon'' = (\tan\delta)\varepsilon' \quad (2)$$

IOP Conf. Series: Materials Science and Engineering **1316** (2024) 012010 doi:10.1088/1757-899X/1316/1/012010

Here ε″ - Imaginary part of dielectric permittivity or dielectric loss factor, tan δ - dielectric loss tangent, ε' - Real part of dielectric permittivity or dielectric constant.

$$\sigma_{ac} = 2\pi f \varepsilon_0 \varepsilon'' \tag{3}$$

Where, σ_{ac} -ac conductivity, f -frequency, ε″ - dielectric loss factor.

Complex impedance components (Z′, Z″) are calculated using following relations:

$$Z' = \frac{\varepsilon''}{\omega\ C_o\left(\varepsilon'^2 + \varepsilon''^2\right)} \tag{4}$$

$$Z'' = \frac{-\varepsilon'}{\omega\ C_o\left(\varepsilon'^2 + \varepsilon''^2\right)} \tag{5}$$

Here. $C_o = \frac{\varepsilon_0 A}{d}$ Here A is the area of electrode plates, and d is the air gap.

3. Results and Discussions

In Figure 1 (a), the X-ray data of the studied samples confirm the formation of single-phase spinel nano-ferrite. The left inset of Figure 1 (a) displays an enlarged view of the 311 peak, revealing a shift of the peak towards relatively lower two theta values, indicating a moderate increase in $a_{exp,}$ as illustrated in the right inset of Figure 1 (a). This indicates that the overall unit cell has expanded slightly. This also suggests the dilation of the spinel unit cell induced by thermal annealing, consistent with the obtained decreasing values of ρ_{XRD} ranging between 5429.58-5424.11 kg/m^3 (See Table 1). In Figure 1 (b), it is apparent with increase in annealing temperature the grain size increases, while the dislocation density slightly decreases (shown as a function of grain size in the inset of figure 1(b), although there is no significant difference between the grain diameters of samples annealed at 500 °C and 600°C (see Table 1). This observed increase in grain size can be attributable to the enhanced migration of grain boundaries and associated intensified diffusion processes at elevated annealing temperatures, supported further by

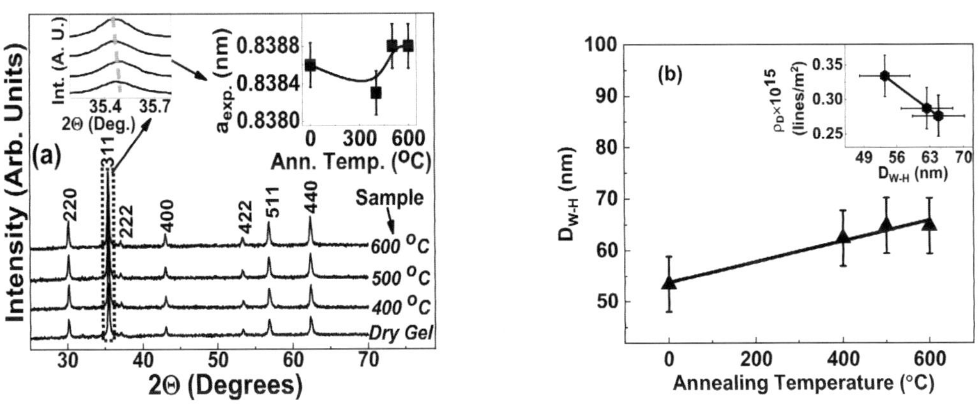

Figure 1. (a) XRD Patterns of $ZnFe_2O_4$ (Left Inset: magnified view of the 311 peak, Right Inset: Variation of $a_{exp.}$ with temperature), **(b)** Dependence of $D_{W\text{-}H}$ on annealing temperature (**Inset:** $D_{W\text{-}H}$ vs dislocation density). Continuous line represents linear fit to the data.

IOP Conf. Series: Materials Science and Engineering **1316** (2024) 012010 doi:10.1088/1757-899X/1316/1/012010

Table 1. Experimental lattice parameter ($a_{exp.}$), Williamson–Hall grain diameter (D_{W-H}), strain (ε), dislocation density (ρ_D), x-ray density (ρ_{XRD}), specific surface area (S), and cell volume (V) of the studied sample $ZnFe_2O_4$.

Samples	a_{exp} (nm) ±0.0002	D_{W-H} (nm) ±1.21	ε ×10^{-3} ±0.2	ρ_D ×10^{15} (Lines/m^2) ±0.031	ρ_{xrd} (kg/m^3) ±5.25	S (m^2/g) ±1.12	V (nm^3) ±0.0004
Dry gel	0.8336	53.43	2.0	0.334	5429.58	20.68	0.5897
Ann. 400°C/2h	0.8383	62.35	2.4	0.287	5435.42	17.70	0.5890
Ann. 500°C/2h	0.8388	64.81	1.8	0.276	5424.69	17.07	0.5902
Ann. 600°C/2h	0.8388	64.74	1.8	0.276	5424.11	17.09	0.5903

the decrease in specific surface area, as depicted in Table 1. With increasing thermal energy, both dislocations, and grain boundaries begin to migrate. It is important to note that dislocations act as obstacles or pinning centers that impede the motion of grain boundaries, thereby hindering further grain growth at higher annealing temperatures.

Table 2 illustrates the thermal annealing-assisted cationic distribution, presenting the calculated, and observed intensity ratios (I_{220}/I_{440}, I_{400}/I_{422}, I_{220}/I_{400}) for the studied samples. Close agreement between the calculated, and observed intensity ratios (see Table 2), confirms the reliability of the computed cationic distribution. Furthermore, Table 2 demonstrates that thermal annealing promotes cationic redistribution, consistent with earlier reports [22]. Examination of Table 2 reveals that Fe^{3+} and Zn^{2+} ions are present on both A and B sites. However, with increasing temperature, there is a decrease in the population of Zn^{2+} ions on the B site, accompanied by a concurrent increase of population of Fe^{3+} ions.

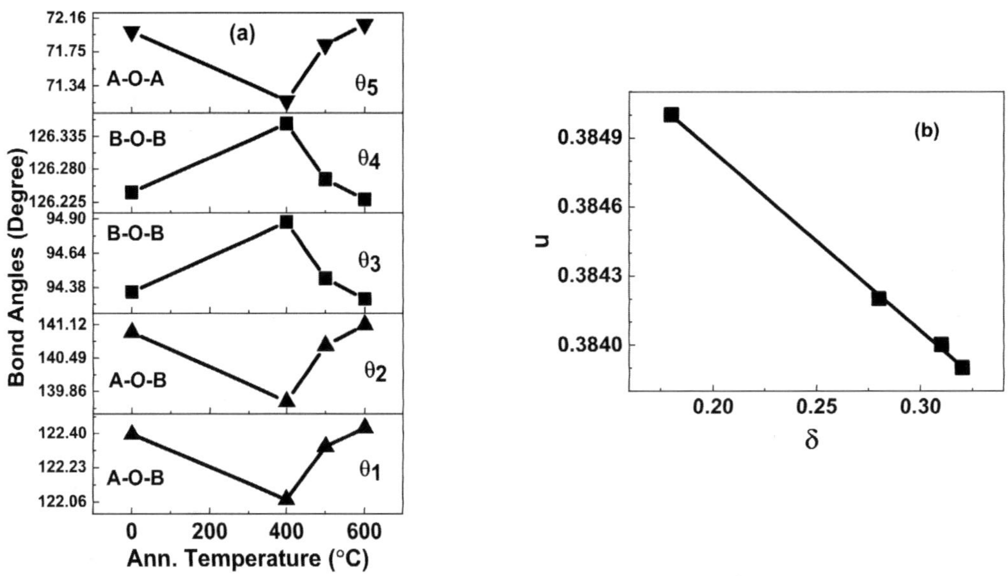

Figure 2. (a) Variation in bond angle with annealing temperature. **(b)** Relationship between oxygen parameter (u) and inversion parameter (δ). The continuous line represents a linear fit to the data.

Table 2. Cationic distribution (for A and B-site), and calculated, observed value of intensity ratios I_{220}/I_{440}, I_{400}/I_{422}, I_{220}/I_{400} for $ZnFe_2O_4$ samples.

Samples $ZnFe_2O_4$	Cation Distribution		I_{220}/I_{440}		I_{400}/I_{422}		I_{220}/I_{400}	
	A site	B site	Calc.	Obs.	Calc.	Obs.	Calc.	Obs.
Dry gel	$(Zn^{2+}_{0.69}$ $Fe^{3+}_{0.31})$	$[Zn^{2+}_{0.31}$ $Fe^{3+}_{1.69}]$	0.8944	0.9960	1.6090	1.6123	1.8827	2.0336
Ann.400°C/2h	$(Zn^{2+}_{0.82}$ $Fe^{3+}_{0.18})$	$[Zn^{2+}_{0.18}$ $Fe^{3+}_{1.82}]$	0.9335	1.1259	1.4689	1.6175	1.9726	2.1017
Ann.500°C/2h	$(Zn^{2+}_{0.72}$ $Fe^{3+}_{0.28})$	$[Zn^{2+}_{0.28}$ $Fe^{3+}_{1.72}]$	0.8911	1.0494	1.5754	1.5055	1.8992	2.2571
Ann.600°C/2h	$(Zn^{2+}_{0.68}$ $Fe^{3+}_{0.32})$	$[Zn^{2+}_{0.32}$ $Fe^{3+}_{1.68}]$	0.97024	1.0150	1.6220	1.5538	1.9441	2.1626

Observed cationic redistribution leads to increase of the inversion degree in the studied samples from 0.18 to 0.32. It is of value to note that bond angles closely relate to the strength of A-O-B, B-O-B, and, A-O-A super-exchange interaction. The bond angles θ_1, and θ_2 provide insight into the A-O-B interaction, while θ_3, and θ_4 offers information on the B-O-B interaction, and θ_5 pertains to the A-O-A super-exchange interaction [1]. Figure 2 (a) illustrates the variation of bond angles with thermal annealing for the studied samples, attributable to the changes in cationic distribution via thermal annealing (see Table 2). Observation of Figure 2 (a) reveals that with increasing thermal annealing temperature, the increase in θ_1, θ_2, and θ_5 suggests the strengthening of the A-O-B, and A-O-A interactions, while the concurrent decrease in θ_3 and θ_4 indicates the weakening of the B-O-B super-exchange interaction. Figure 2 (b) exhibits a linear relationship (fitted using the equation $u = 0.3864 - 0.0078\ \delta$) between the oxygen parameter (u), and the inversion parameter (δ), consistent with previous reports [23]. For the samples under study, u ranges from 0.3839 to 0.3850, surpassing its ideal value of 0.375 [24], indicating a higher degree of disorder compared to the ideal spinel structure.

Figure 3 illustrates the variation of the Néel magnetic moment ($M_{s(th)}$ at 0 K) with annealing treatment. The observed behaviour of the Néel magnetic moment ($M_{s(th)}$ at 0 K) is attributed to the strengthening of the A-O-B, and A-O-A interactions, accompanied by the corresponding weakening of the B-O-B super-exchange interaction. The relatively higher value of the Néel magnetic moment for the sample annealed at 400 °C can be ascribable to a significant difference in magnetic moment between

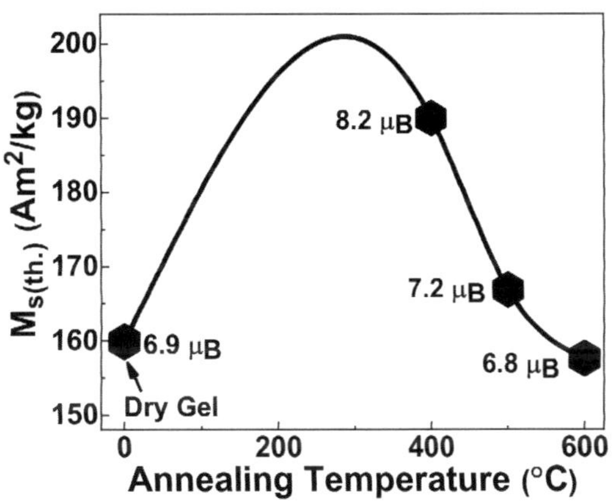

Figure 3. Variation of $Ms_{(th)}$ at 0 K with respect to thermal annealing, along with the corresponding Néel magnetic moment values (in μ_B), is also shown. Lines connecting the points aid visualization.

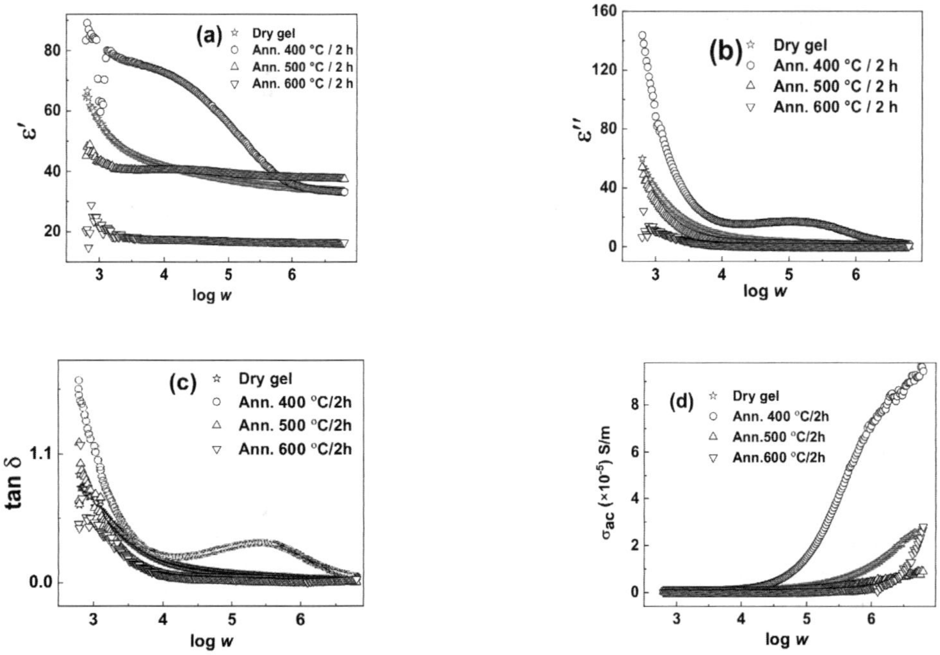

Figure 4. The thermal annealing temperature dependence of the **(a)** dielectric constant (ε'), **(b)** dielectric loss (ε''), **(c)** tangent loss (*tan δ*) and **(d)** ac conductivity (σ_{ac}), as a function of frequency.

A, B sites, which, suggests an increasing degree of disorder, as indicated by the observed higher values of the oxygen parameter. This can be understood in terms of the maximum number of Fe^{3+} ions occupying the B site compared to the maximum number of Zn^{2+} ions residing on the A site (see Table 2). Further decrease of the Néel magnetic moment at higher annealing temperatures is due to an increase in the number of Zn^{2+} ions on the B-site (see Table 2).

Figures 4 (a-d) respectively illustrate the room temperature frequency (range: 100 Hz – 1MHz)-dependent complex dielectric constant components (ε', ε''), tangent loss (tan δ), and AC conductivity (σ_{ac}) of the studied $ZnFe_2O_4$ nanoferrite samples. Dielectric properties for the studied samples appear to exhibit significant alterations with annealing treatment. It can be observed from figure 4 (a) that the dielectric constant (ε') is higher in the low-frequency region for as burnt as well as annealed samples. However, as the frequency increases, the value of ε' decreases due to the electric dipoles inability to respond to the changing applied AC electric field. Furthermore, it is of value to note that the magnitude of the dielectric constant increases with increasing annealing temperature. This may be occurring due to the higher temperatures, which facilitate the alignment of more dipoles along the electric field, consequently resulting in an increase in the dielectric constant [25]. Moreover, the variation of dielectric constant (ε') with frequency can be explained by the Maxwell-Wagner two-layer model for space charge or interfacial polarization, in conjunction with Koop's phenomenological theory [26-28]. According to this model, dielectric materials are composed of numerous well-conducting grains separated by poorly conducting thin grain boundaries. Upon application of an external electric field, the charge carriers can easily migrate from the grains, and accumulate at the grain boundaries. This phenomenon leads to the significant polarization, and hence a high dielectric constant. In dielectric materials, dielectric losses typically occur due to the absorption of electrical energy, which is utilized for the rotations of dipolar molecules [29]. Figure 4 (b and c) suggests that the complex component of dielectric constant ε'', and dielectric loss tangent tan δ for the sample annealed at 400 °C are relatively higher. It may be ascribable

Table 3. Parameters obtained by fitting σ_{ac} data using Joncher's power law.

Samples	*s*	*A* ($Sm^{-1}rad^{-n}$)
Dry gel	0.5361	1.8324×10^{-5}
Ann. 400°C/2h	0.6720	1.0877×10^{-5}
Ann. 500°C/2h	0.4618	1.5221×10^{-5}
Ann. 600°C/2h	0.5762	1.7021×10^{-6}

to higher in-homogeneities that can create an absorption current due to the space charge formation that follow in a dielectric loss as was also reported earlier in case of nano-phase materials [30] This is in conjunction with cationic redistribution (see table 2) ensuing maximum migration of Fe^{3+} and Zn^{2+} ions towards B and A sites respectively. Figure 4 (b and c) suggest that complex component of dielectric constant ε" and dielectric loss tangent tan δ exhibited nearly identical variation with frequency. It should be noted that a relaxation peak (broad hump or maximum) also occur for both at certain frequency value in increasing frequency regime. It seems to have occurred due to the condition when the hopping frequency of charge carriers matches the frequency of the applied electric field [31]. When the hopping frequency of charge carriers between the ions exceeds that of the applied electric field, the charge carriers can track the field, leading to increased absorption of the propagating electromagnetic waves, and consequently higher losses at low frequencies. Conversely, when the hopping frequency of charge carriers is lower than that of the applied field, the charge carriers cannot keep pace, resulting in negligible absorption, and lower losses observed at higher frequencies. Additionally, losses originating from the dissipation of energy due to relaxation of dipoles also adds up to dielectric loss tangent (tan δ) [32]. The figure 4 (d) illustrates the variation of AC conductivity (σ_{ac}) with frequency. It can be observed that the AC conductivity (σ_{ac}) stabilizes at a nearly constant value at lower frequencies. The increase in conductivity at higher frequencies can be attributable to the movement of free, and bound charges. In the high-frequency region, the conductivity mechanism is controlled by charge carriers tunnelling, while at low frequencies, it mainly relies on the slow motion of charge carriers [26, 27], through conducting grains, and the poorly conducting (dielectric) grain boundaries. Grain boundaries exhibit high resistance at low frequencies, and low resistance at high frequencies, consequently enhancing the conductivity [31]. It is noteworthy that the frequency dependence of conductivity is described by a simple expression given by- $\sigma_{total} = \sigma_{dc} + Aw^s$, where the second term is representing the frequency dependent ac conductivity also known as Jonscher's power law [33]. This frequency dependent ac conductivity can also be expressed as per the following relation: $\log \sigma_{ac} = \log A + s \, log \, w$. Table 3 summarize the value of parameters s and A. These parameters are a result of a linear fit to the log-log variation of ac conductivity versus frequency. Here, A represents the strength of polarizability, and s is a dimensionless parameter that is dependent on temperature. Variation of s, and A with temperature determines the conduction mechanism. It is of value to note that for the correlated barrier hopping (CBH) mechanism this parameter s decreases with temperature [34]. Perusal of table 3 suggests that the value of $s < 1$ which is an indicative of the hopping due to translational motion [35], and it appears that a correlated barrier hopping remains dominant for the present studied samples/system.

Figure 5(a) illustrates Cole-Cole plots, depicting the variation of complex permittivity (imaginary part (ε″), and real part (ε′) of dielectric constant) for the dry-gel, and the inset of this figure shows the Cole – Cole plots for the samples annealed for 2 hours at 400 °C, 500 °C, and 600 °C. It is worth noting that the real component of complex permittivity (ε′) signifies the material's dielectric constant, whereas the imaginary component denotes losses stemming from the absorption of electric energy. Observing Figure 5 (a), a linear type behaviour is evident at lower frequencies, while for higher frequencies; the behaviour gradually shifts, resembling that of a semi-circular arc. This shift appears to arise from electrical conduction by bulk grains. [26, 28, 36]. The changes in the components of the complex dielectric constant shown in Figure 5(a) imply primarily a rise in absorption losses as the dielectric constant increases initially at lower frequencies. It appears that the influence of grain boundaries

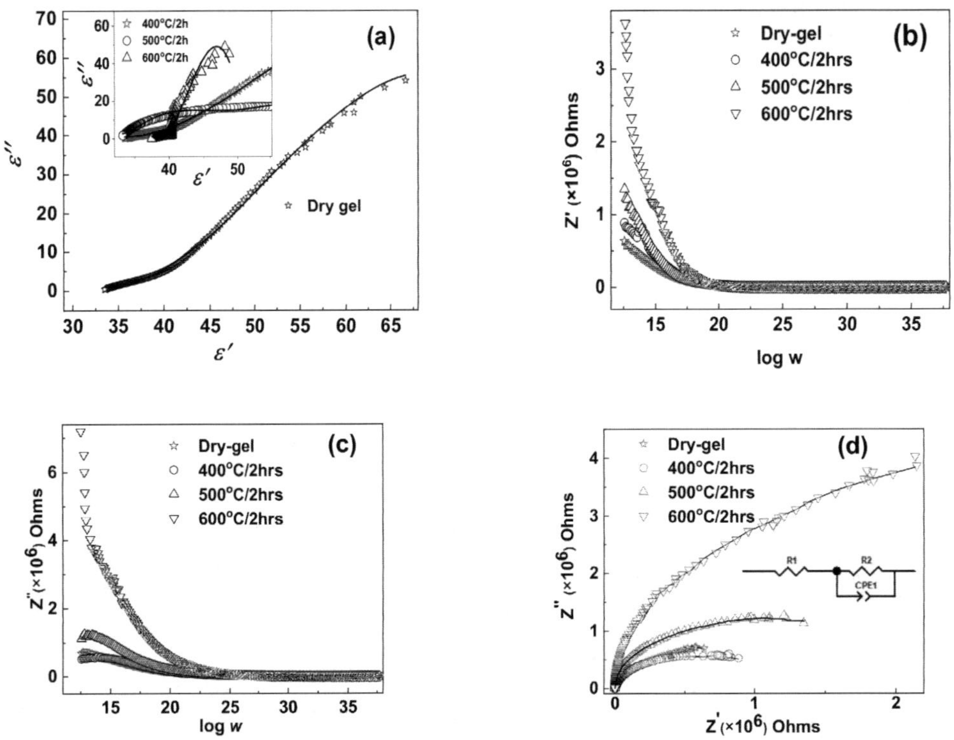

Figure 5 (a) Cole-Cole plot between imaginary part (ε″) and real part (ε′) of dielectric constant for as burnt $ZnFe_2O_4$: Inset shows the same for annealed samples (**b**) Real part of impedance versus log w **(c)** Imaginary part of impedance versus log **w (d)** Nyquist -plot for the dry-gel and annealed samples (Inset shows the equivalent circuit.

become pronounce over time, due to their lower permittivity, and conductivity compared to the grains. As the annealing temperature increases, the grain size also increases. Consequently, while this may reduce the overall grain boundary region, cationic redistribution simultaneously enhances the contribution of the grain boundary region to the observed dielectric response by affecting the interfacial polarization, attributed to the increased heterogeneity of the permittivity. Usually, complex impedance is measured experimentally. Nevertheless, in our case, we have deduced the impedance spectra using equations (4) and (5), which establishes a connection between the complex impedance components (Z', Z"), and the respective components of complex permittivity. Figure 5 (b) shows the variation of real part of impedance (Z') as a function of angular frequency. From the figure 5 (b) it can be seen that corresponding to all annealing temperatures the Z' decreases rapidly as frequency increases and becomes almost frequency independent for higher frequencies and seems to exhibit almost similar values for all samples owing to release of trapped charges. More or less similar trend is displayed in terms of change of imaginary part of impedance (Z") with angular frequency that can be seen in figure 5 (c). Rapid decrease of Z" for higher annealing temperature suggests increased hopping of localized charge carriers [37, 38].

Figure 5 (d) represents the Nyquist plots showing the variation of Z''versus Z' for dry-gel, and annealed samples. The plots suggests a dispersive type behaviour rather than a centered semicircle on the real axis, indicating a relaxation behaviour that is kind of non-Debye type. The figure 5 (d) delineates two regions: the grain region at higher frequency and the grain boundary region at lower frequency.

The curves were fitted using Zview software, taking an equivalent circuit consisting of a resistive element in series with a parallel resistance-constant phase element (R-CPE) circuit shown in the inset of figure 5 (d). The Constant Phase Element, added into the circuit, is to account for the grain boundary properties stemming from non-ideal capacitive behaviour, possibly due to the existence of multiple relaxation processes [39, 40]. Perusal of figure 5 (b) also suggests that with increasing annealing temperature there is an increase in the rate of thermally activated electron hopping due to the release of trapped charge carriers indicating a decrease in defects or impurities within the growing grains. This is also associated with polarization of charges due to the build-up of charge at the interfacial grain boundary region [40].

4. Conclusions

We examined the structural, and dielectric characteristics of $ZnFe_2O_4$ nano-ferrites, prepared via sol-gel auto-combustion, with grain sizes ranging between 53.43 nm to 64.81 nm using XRD, and dielectric analyses. Thermal annealing led to an increased lattice parameter ($a_{exp.}$), a rise in Zn^{2+} ion population, and a decrease in Fe^{3+} ions on the B-site. This altered the inversion degree (0.18 to 0.32), and strengthened the A-O-B and A-O-A interactions while weakening the B-O-B super-exchange interaction. Theoretical magnetization at 0 K ($M_{s(th)}$) also decreased. Dielectric properties changed with annealing, showing a decline in both real (ε') and imaginary (ε") components of the dielectric constant with increasing frequency. The loss tangent (tan δ) decreased consistently with frequency, hitting a low point for the sample annealed at 600 °C for 2 hours. AC conductivity (σ_{ac}) significantly increased at higher frequencies in the sample annealed at 400 °C for 2 hours, hinting at a correlated barrier hopping mechanism. These results underscore the influence of annealing-induced cationic redistribution on dielectric properties of $ZnFe_2O_4$. The complex impedance plots further highlighted the rising impact of grain boundaries on these properties, suggesting the studied samples potential for high-frequency device applications.

Acknowledgments

Authors thank Dr. R. J. Chaudhary, UGC-DAE CSR, Indore, for XRD measurements. This work is supported by Seed money grant: No. Dev/Seedmoney2.0/2020-21/649, 20 Jan. 2022 of D. A. University, Indore (India).

References

[1] Smit J and Wijn H P J 1959 *Ferites* (Eindhoven, The Netherlands: Philips Technical Library)
[2] Kane S N, Verma R, Tiwari P and Mazaleyrat F 2019 *AIP Conf. Proc.* **2142** 020003-1-8
[3] Verma R, Kane S N, Tiwari P and Mazaleyrat F 2019 *AIP Conf. Proc.* **2142** 160001-1–160001-5
[4] Kane S N, Verma R, Modak S S, Reddy V R and Mazaleyrat F 2023 *Hyperfine Interactions* **244** 3-1-15
[5] Parmar C, Verma R, Modak S S, Mazaleyrat F and Kane S N 2022 *ECS J. of Solid State Sci. and Tech.* **11** 053015-1-9
[6] Verma R, Modak S S, Deshpande U P and Kane S N 2020 *AIP Conf. Proc.***2352** 040018-1-5
[7] Verma R, Modak S S, Ghosh A and Kane S N 2020 *AIP Conf. Proc.* **2352** 040006-1-5
[8] Parmar C, Verma R, Modak S S, Mazaleyrat F, Tatarchuk T and Kane S N 2023 *Mol. Cryst. Liq. Cryst.* **767** 115–125
[9] Verma R, Kane S N, Tiwari P, Modak S S, Tatarchuk T and Mazaleyrat F 2018 *Mol. Cryst. Liq. Cryst* **674** 130-141
[10] Low Kin O and Sale Frank R 2003 *J. Magn. Magn Mater* **256** 221–226
[11] Goev G, Masheva V, Ilkov L, Nihtianova D and Mikhov M 2003 *Proceedings of the fifth General Conference of the Balkan Physical Union BPU-5* 687.
[12] Ajmal M and Maqsood A 2008 *J. Alloys Compd.* **460** 54–59
[13] Jadhav S A 2000 *Mater. Chem. Phys.* **65** 120–123

[14] Parmar C, Verma R, Modak S S, Mazaleyrat F, Tatarchuk T and Kane S N 2023 *Mol. Cryst. Liq. Cryst.* **767** 115–125
[15] Yadav R S, Kuřitka I, Vilcakova J, Havlica J, Masilko J, Kalina L, Tkacz J, Švec J, Enev V and Hajdúchová M 2017 *Adv. Nat. Sci.: Nanosci. Nanotechnol* **8** 045002-1-14
[16] Verma R, Kane S N, Deshpande U P and Mazaleyrat F 2021 *Mater. Today: Proc.* **46** 2205–2211
[17] Verma R, Kane S N, Tiwari P, Modak S S and Mazaleyrat F 2019 *AIP Conference Proceedings* **2142** 160015-1-5
[18] Bertaut, E F 1950 Comptes Rendus Hebdomadaires des Seances de I'Academie des Sciences **230** 213–215
[19] Tanna A R and Joshi H H 2013 *International Journal of Physical and Mathematical Sciences* **7** 334–341
[20] Pardavi-Horvath M 2000 *J. Magn. Magn. Mater.* **215-216** 171-183
[21] Rekaby M 2020 *Applied Physics A* **126** (8), 664-1-9
[22] Kane S N, Raghuvanshi S, Satalkar M, Reddy V R, Deshpande U P, Tatarchuk T R and Mazaleyrat F 2018 *AIP Conf. Proc.* **1953** 030089-1-5
[23] Bindu K, Ajith K M, Nagaraja H S 2019 *Mater. Res. Express* **6** 045011.
[24] Tiwari P, Kane S N, Verma R, Tatarchuk T and Mazaleyrat F (Chapter 29) Nanochemistry, Biotechnology, Nanomaterials, and Their Applications. In: Fesenko, O., Yatsenko, L. (eds.), Nanocomposites, Nanostructures, and Their Applications 222, pp. 431–442. Springer Proceedings in Physics, Springer International Publ. Co., part of Springer Nature (2019).
[25] Shakir M, Singh B K, Gaur R K, Kumar B, Bhagavannarayana G and Wahab M A 2009 *Chalcogenide Lett.* **6** 655
[26] Wagner K W 1913 *Ann. Phys.* **40** 817–819
[27] Koops C G 1951 *Phys. Rev.* **83** 121-124
[28] Jaita P, Watcharapasorn A, Kumar N, Cann D P and Jiansirisomboon S 2015 *Elect. Mater. Lett.* **11** 828–835
[29] Prabaharan D, Sadaiyandi K, Mahendran M and Sagadevan S 2016 *Mat. Res.* **19** 478–482
[30] S Suresh 2013 *Int. J. Phy. Sci.* **8** 1121–1127
[31] Murugesan C, Okrasa L and Chandrasekaran G 2017 *J Mater Sci: Mater Electron* **28** 13168–13175
[32] Khirade P P, Chavan A R, Somvanshi S B, Kounsalye J S and Jadhav K M 2020 *Mater. Res. Express* **7** 116102-1-17
[33] Jonscher A K 1997 *Nature* **267** 673
[34] Gopalan E V, Mallini K A, Saravanan S, Kumar D S, Yoshida Y and Anantharaman M R 2008 *J. Phys. D Appl. Phys.* **41** 185005
[35] Funke K 1993 *Prog. Sol. State Chem.* **22** 111
[36] Kambale R C, Shaikh P A, Bhosale C H, Rajpure K Yand Kolekar Y D 2009 *Smart Mater. Struct.* **18** 085014-1-6
[37] Costa M M, Pires J G F M, Terrezo A J, Graca M P F and Somba A S B 2011 *J. Appl. Phys.* **110** 034107
[38] Barsoukov E and Macdonald J R 2005 *Impedance Spectroscopy: Theory, Experiments and Applications* (Wiley, New York)
[39] Younas M, Nadem M, Atif M and Grossinger R 2011 *J. Appl. Phys.* **109** 093704
[40] Ali H, Karim S, Rafiq M .A, Maaz K, Rahman A U, Nisar A and Ahmad M 2014 *J. Alloys Comp.* **612** 64–68

IOP Conf. Series: Materials Science and Engineering **1316** (2024) 012011 doi:10.1088/1757-899X/1316/1/012011

Absorption and emission maxima prediction of fluorescent organic materials using optimized neural network models

Kapil Dev Mahato*[1]

[1] Department of Physics, National Institute of Technology, Jamshedpur, Jharkhand, Pin-831014, India.

E-mails: kapildevmahato4@gmail.com, 2018rsphy005@nitjsr.ac.in

Abstract. The aim of the study is to predict the absorption and emission maxima from the given dataset, which consists of 3066 fluorescent organic materials. To fulfill this requirement, five optimized neural network models are employed. Out of them, the wide neural network (WNN) model outperformed the other models on validation as well as test results. The results were obtained on the basis of three evaluation parameters: mean absolute error (MAE), root mean squared error (RMSE), and the coefficient of determination (R^2). According to the validation scores, the WNN was the best-predictive model with MAE-18.05 nm, RMSE-28.93 nm, and R^2-89.55%, and MAE-29.58 nm, RMSE-42.62 nm, and R^2-78.87% for absorption and emission maxima, respectively. On the other hand, on the basis of tested scores, the WNN was the best-predictive model with MAE-19.35 nm, RMSE-29.33 nm, and R^2-92.14%, and MAE-29.17 nm, RMSE-41.87 nm, and R^2-79.46% for absorption and emission maxima, respectively. The presented automated method does not require an extensive understanding of computer programming to estimate the absorption and emission maxima. The absorption and emission maxima may be predicted, which helps with the design of new fluorescent organic materials and their many uses in electronics, chemistry, materials science, medicine, and other areas.

1. Introduction

Fluorescent organic materials are widely explored in the design and discovery of new materials, photovoltaic cells, light sensors, imaging applications, energy harvesting applications, and so on [1,2]. The majority of applications necessitate an understanding of the absorption and emission maxima spectra of fluorescent organic materials in order to be appropriately evaluated prior to usage. Numerous endeavours' have been undertaken to synthesize fluorescent organic materials possessing the intended photophysical characteristics [3]. Finding the absorption and emission maxima spectra of the chosen fluorescent organic materials is a difficult problem for experimentalists with a short period of time and less expenditure. To solve this problem using optimized neural network models before the experiment starts, this approach helps researchers obtain prior information about the fluorescent materials. To accomplish this work, 3066 individuals' fluorescent organic materials are taken into account for training the applied machine learning models. Here, five machine learning (ML) regression models are considered to predict the absorption and emission maxima of fluorescent materials. All implemented regression models are neural network (NN)-based, such as Narrow NN (NNN), Medium NN (MNN), Wide NN (WNN), Bilayered NN (BNN), and Trilayered NN (TNN). Before applying these models, the SMILES (Simplified Molecular Input Line Entry System) of materials and solvents is converted into machine-readable information using Morgan's fingerprint to establish relations among the attributes, consequently enhancing the models' performances.

In the past few years, Density Functional Theory (DFT), Time-Dependent DFT (TD-DFT), and ab initio methods have been used for designing new fluorescent organic materials for specific applications [4]. Despite being able to accurately anticipate the emission and absorption maxima of fluorescent organic materials, these approaches are more expensive and require more time when doing large-scale molecular structure assessments. The fastest computational approaches, such as ML, deep learning (DL), and artificial intelligence (AI), are reliable and efficient tools for promptly and precisely estimating the emission maxima and absorption maxima of fluorescent organic materials. Albeit, there have been limited related publications reported using ML in this direction, such as absorption maxima [5–8] and emission maxima [8–11]. Recently, N. Tomar and her colleagues published an article and predicted the absorption maxima of fluorescent organic materials (dyes) using an artificial NN (ANN) model [4]. The above ML approaches require programming knowledge to perform specific jobs. However, the proposed optimized neural network models do not require much programming knowledge. The optimized models have an easy-to-use interface and automate steps during the experiment. There is a lack of studies to predict the absorption and emission maxima of fluorescent organic materials using optimized neural network models.

In the current work, five optimized NN models have been employed to investigate the given dataset (3066 records) for estimating the absorption and emission maxima. The models' performances are assessed by the three statistical parameters: MAE, RMSE, and R^2. Using these parameters to calculate validation (train) and test results. The present study conducted a comparative analysis among the employed models and achieved the best predictive neural network model among them. This study considers 91% of the total data for training and the rest of the data for testing the investigated models. Before the experiment, string data attributes were decoded by the Morgan fingerprint and proceeded through ten-fold cross-validation techniques [12]. According to the validation scores, the WNN is the best-predictive model with MAE-18.05 nm, RMSE-28.93 nm, and R^2-89.55%, and MAE-29.58 nm, RMSE-42.62 nm, and R^2-78.87% for absorption and emission maxima, respectively. On the other hand, on the basis of tested scores, the WNN is the best-predictive model with MAE-19.35 nm, RMSE-29.33 nm, and R^2-92.14%, and MAE-29.17 nm, RMSE-41.87 nm, and R^2-79.46% for absorption and emission maxima, respectively. Based on the analysis, the optimized WNN predictive model is suitable for predicting the absorption and emission maxima of fluorescent organic materials with robust extrapolation capabilities. The proposed models are available under the MATLAB Regression Learner App, which has some limitations. The app's flexibility is restricted when dealing with huge and complex datasets. In addition, no new hybrid ensemble models can be constructed from the proposed models. Moreover, model deployment to the MATLAB production server via MATLAB online is not supported. Python-based neural network models are being explored as a potential solution to these issues, but they are not perfect as their effectiveness depends on the volume of data used in the research.

2. Materials and methods

2.1. Dataset construction

A dataset with the maximum absorption and emission wavelengths of fluorescent organic materials was constructed in 2021 by Ju et al. [13]. There were 11 attributes and 4386 instances in this dataset. Five out of the eleven attributes were solvatochromic parameters (acidity (SA), basicity (SB), polarizability (SP), dipolarity (SdP), and polarity (Et30)); three were three photophysical properties parameters (quantum yield, absorption, and emission maxima); one was the materials' SMILES; and one was the reference for the materials' gathered data. The remaining feature was solvents in SMILES format. This study employed the Morgan fingerprint [14,15], which is available under the RDKit library [16]. This is a kind of molecular fingerprint that can be acquired without the need for quantum-mechanical calculations. It was used in this work to decode the SMILES of fluorescent organic materials and solvents. Omitting the missing and duplicate values from the given dataset was necessary for smoothly conducting the real-time experiment in the MATLAB machine learning regression app. After the data cleaning, the examined dataset was divided into two parts: training (91%) and testing (9%), which recorded 3066 individual fluorescent materials. The examined dataset was proceeded through ten-fold cross-validation techniques. This technique helps to provide a more robust estimate of the model's performance compared to a single train-test split. It ensures that the model is evaluated on different subsets of the data, helping to reduce the impact of the specific split on the model's performance evaluation.

2.2. *Neural network models*

Figure 1 shows the architecture of optimized NN models. These are made up of interconnected nodes, or artificial neurons, grouped into layers and ReLU activation functions that add nonlinearity to the network. This non-linearity enables the neural network to capture deeper relationships within the data by combining linear operations (weights and biases) and non-linear activations, allowing the network to learn and represent complex functions. The weights in the first layer encode information about the input data, converting it to a more abstract and intelligible representation. These encoded features are then passed on to the next layers for further refinement and abstraction. The second and succeeding layers expand on these features, capturing more intricate interactions and tailoring the representation for the specific task at hand. The depth and layout of layers in a NN influence the model's ability to learn and generalize from input data.

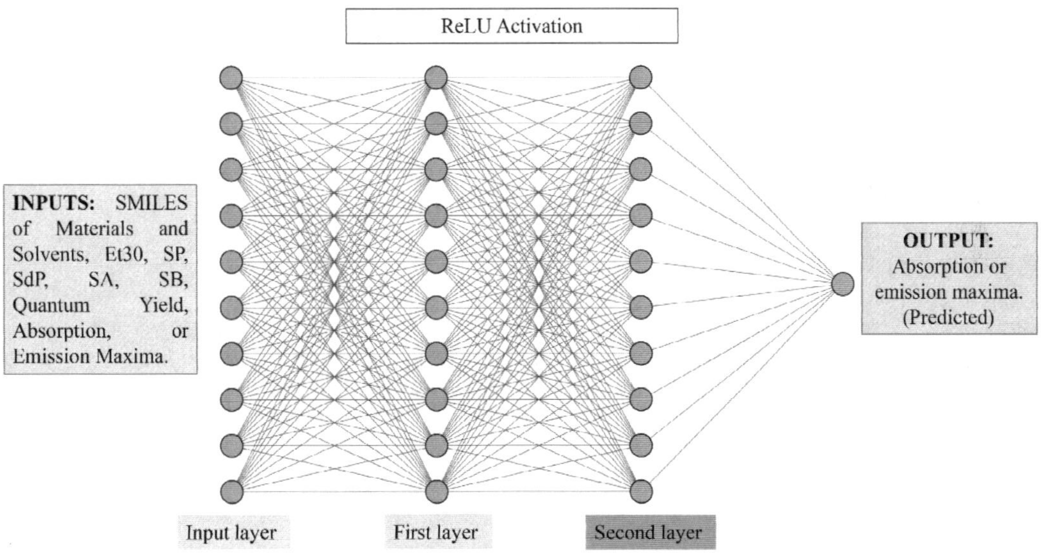

Figure 1. Architecture of optimized neural network models

The models NNN, MNN, WNN, BNN, and TNN refer to different configurations or architectural characteristics of NNs used for regression tasks.

An NN with fewer neurons or units in its hidden layers is referred to as an "NNN" in most cases. There may be fewer hidden layers or neurons per layer in the network layout. An NN can be computationally effective and less susceptible to overfitting because it is very simple and has a smaller number of parameters. However, it might not be able to represent complicated patterns well and would need more training data to work well. If we increase the number of neurons compared to an NNN, a developed NN is referred to as an "MNN" if its hidden layers include a manageable number of neurons. It strikes a compromise between a vast network's complexity and a tight network's simplicity. An average number of hidden layers, or neurons per layer, may be present in a medium network architecture.

It offers a fair mix between model complexity and computing performance and can capture a wider range of patterns. If we increase the large number of neurons or units in the hidden layer, then it is called a "WNN", which is capable of capturing complex and intricate patterns in the data. It has a strong learning capacity and is likely to yield improved prediction results. However, these networks may require more computational resources and may be more prone to overfitting, especially when the training data is limited.

Another approach to NNs is the hidden layer approach. Suppose we increase the hidden layer into two and three layers, which are called BNN and TNN, respectively. The bilayered architecture can capture non-linear relationships between the input features and the target variable. Compared to a single-layer network, it offers a higher level of complexity, enabling more adaptable modeling. As opposed to the bilayered network, the trilayered architecture includes a further hidden layer, allowing for even more intricate modeling. This extra layer can identify complex linkages and patterns in the data. However, it also results in an increase in the number of parameters, which necessitates additional training data and computer power [17].

2.3. *Evaluation parameters*

The evaluation parameters were MAE, RMSE, and R^2 for calculating the employed regression models.

$$MAE = \frac{1}{n}\sum_{i=1}^{n}|\hat{y}_i - y_i| \quad (1)$$

$$RMSE = \sqrt{\frac{1}{n}\sum_{i=1}^{n}(\hat{y}_i - y_i)^2} \quad (2)$$

$$R^2 = 1 - \frac{\sum_{i=1}^{n}(\hat{y}_i - y_i)^2}{\sum_{i=1}^{n}(\hat{y}_i - \bar{y})^2} \quad (3)$$

Where, y, $\hat{y}$, $\bar{y}$ are the true response, predicted response, and the average of the predicted response, respectively. 'n' is the number of data points.

3. Results and discussions

This study proposed five neural network models (NNN, MNN, WNN, BNN, and TNN) to predict absorption and emission wavelength maxima in the examined dataset. The dataset was evaluated using three statistical parameters, namely MAE, RMSE, and R^2, and the best models among the implemented neural network models were found. All models passed through ten-fold cross-validations and obtained validation results on trained data (91%). Further, the model was tested using test-size data (9%). Before the implementation of optimized neural network models, the SMILES of material and solvent information were transferred into a computer-readable format using the Morgan fingerprint molecular descriptor. This fingerprint is able to manifest the key structural feature of molecular skeletons and functional groups, which is directly associated with the photophysical properties of fluorescent materials. Besides, the photophysical properties are also dependent on solvents, particularly for molecules with intermolecular charge transfer and spin-orbit coupling during interactions. The photophysical properties of fluorescent organic materials were influenced by the solvatochromic parameters (SA, SB, SP, SdP, and Et30) of solvents [18,19]. That's why these parameters were taken into account in the examined dataset.

3.1. *Absorption maxima prediction*

Here, five neural network models were applied to calculate the validation and test score of absorption maxima prediction, and their corresponding outcomes are tabulated in Table 1. The WNN model was the best predictive model in both condition validation and test scores. The WNN model validation scores were MAE-18.05 nm, RMSE-28.93 nm, and R^2-89.55%, while the WNN test scores were MAE-19.35 nm, RMSE-29.33 nm, and R^2-92.14%. Five proposed models (NNN, MNN, WNN, BNN, and TNN) obtained MAEs of 21.89 nm, 21.38 nm, 18.05 nm, 22.18 nm, and 22.21 nm for validation scores and 25.22 nm, 24.22 nm, 19.35 nm, 28.25 nm, and 25.34 nm for test scores, respectively (Figure 2a). NNN,

MNN, WNN, BNN, and TNN obtained RMSEs of 39.61 nm, 35.87 nm, 28.93 nm, 40.31 nm, and 40.93 nm for validation scores and 43.61 nm, 37.85 nm, 29.33 nm, 49.93 nm, and 44.82 nm for test scores, respectively (Figure 2b). In terms of accuracy, five proposed models (NNN, MNN, WNN, BNN, and TNN) obtained R^2 values of 80.40%, 83.93%, 89.55%, 79.71%, and 79.08% for validation scores, and 82.62%, 86.91%, 92.14%, 77.21%, and 81.64% for test scores, respectively (Figure 2c).

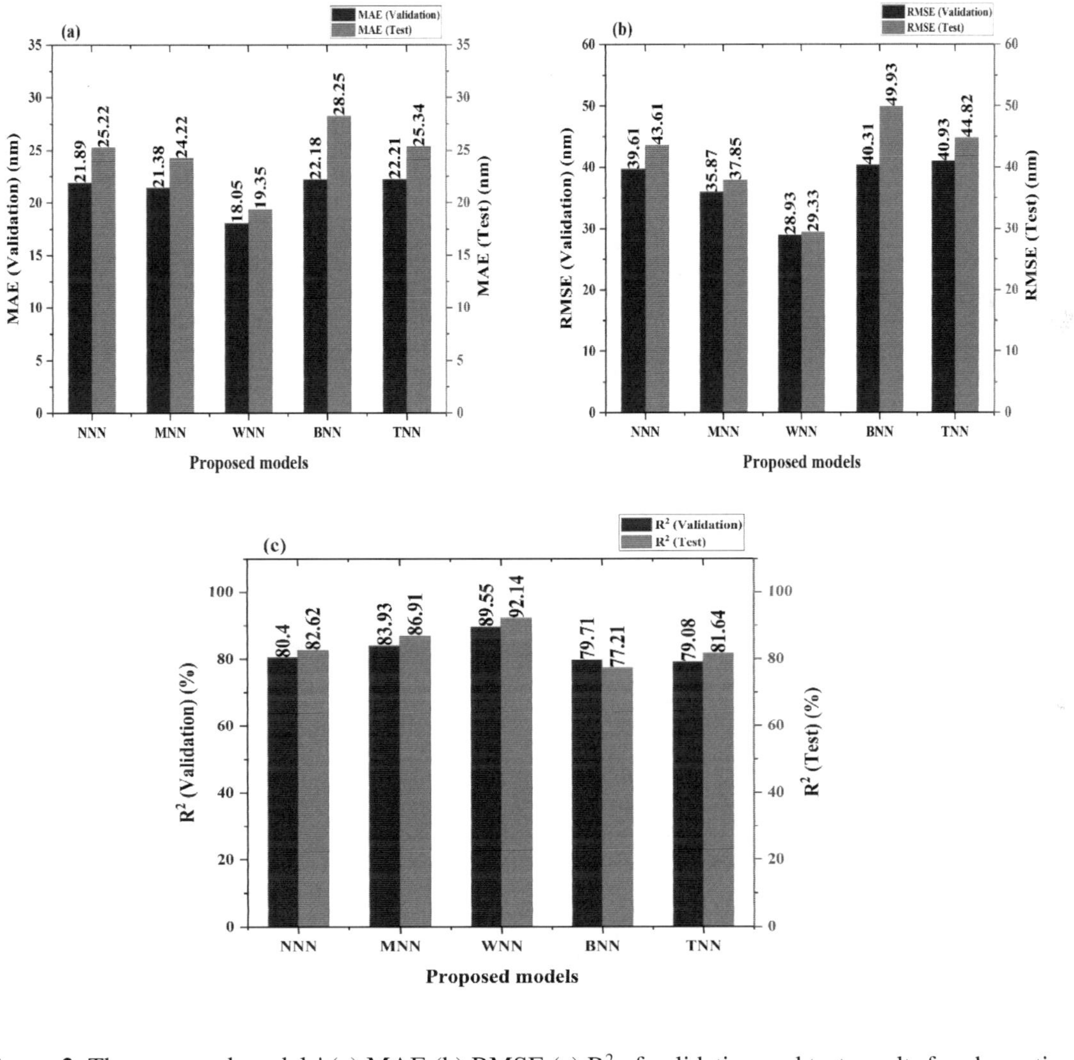

Figure 2. The proposed models' (a) MAE (b) RMSE (c) R^2 of validation and test results for absorption maxima prediction.

When comparing the implemented models, in terms of validation scores, three models (NNN, MNN, and WNN) achieved above 80% accuracy, and two models (BNN and TNN) achieved less than 80% accuracy. The TNN model had the lowest accuracy of 79.08% among the implemented models. Whereas in terms of test scores, four models (NNN, MNN, WNN, and TNN) got satisfactory results above 80% accuracy, and the one remaining model (BNN) got less than 80% accuracy. The WNN outperformed the other models in both validation (train) and test results, and their predicted response versus true response results are graphically illustrated in Figure 3.

Table 1. Comparative study of proposed models on (a) MAE, (b) RMSE, and (c) R^2 in both validation and test results for absorption maxima prediction.

Sl. No.	Proposed models	MAE (Validation) (nm)	RMSE (Validation) (nm)	R^2 (Validation) (%)	MAE (Test) (nm)	RMSE (Test) (nm)	R^2 (Test) (%)
1	NNN	21.89	39.61	80.40	25.22	43.61	82.62
2	MNN	21.38	35.87	83.93	24.22	37.85	86.91
3	WNN	18.05	28.93	89.55	19.35	29.33	92.14
4	BNN	22.18	40.31	79.71	28.25	49.93	77.21
5	TNN	22.21	40.93	79.08	25.34	44.82	81.64

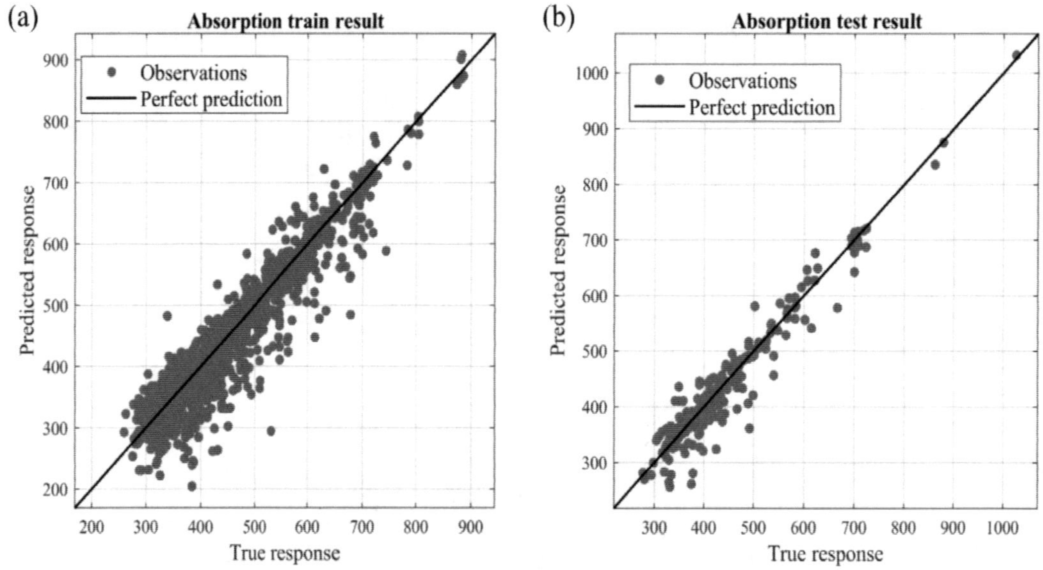

Figure 3. Correlation plots of absorption maxima on (a) validation (train) and (b) test results of the WNN model.

3.2. Emission maxima prediction

Here, five neural network models were applied to calculate the validation and test score of emission maxima prediction, and their corresponding outcomes are tabulated in Table 2. The WNN model was the best predictive model in both condition validation and test scores. The WNN model validation scores were MAE-29.58 nm, RMSE-42.62 nm, and R^2-78.87%, while the WNN test scores were MAE-29.17 nm, RMSE-41.87 nm, and R^2-79.46%. Five proposed models (NNN, MNN, WNN, BNN, and TNN) obtained MAEs of 33.02 nm, 30.98 nm, 29.58 nm, 32.84 nm, and 32.30 nm for validation scores and 33.33 nm, 29.16 nm, 29.17 nm, 29.30 nm, and 33.53 nm for test scores, respectively (Figure 4a). NNN, MNN, WNN, BNN, and TNN obtained RMSEs of 51.11 nm, 44.51 nm, 42.62 nm, 51.97 nm, and 52.33 nm for validation scores and 50.02 nm, 42.53 nm, 41.87 nm, 43.71 nm, and 52.85 nm for test scores, respectively (Figure 4b). In terms of accuracy, five proposed models (NNN, MNN, WNN, BNN, and TNN) obtained R^2 values of 69.62%, 76.95%, 78.87%, 68.58%, and 68.15% for validation scores, and 70.69%, 78.81%, 79.46%, 77.62%, and 67.28% for test scores, respectively (Figure 4c).

IOP Conf. Series: Materials Science and Engineering **1316** (2024) 012011 doi:10.1088/1757-899X/1316/1/012011

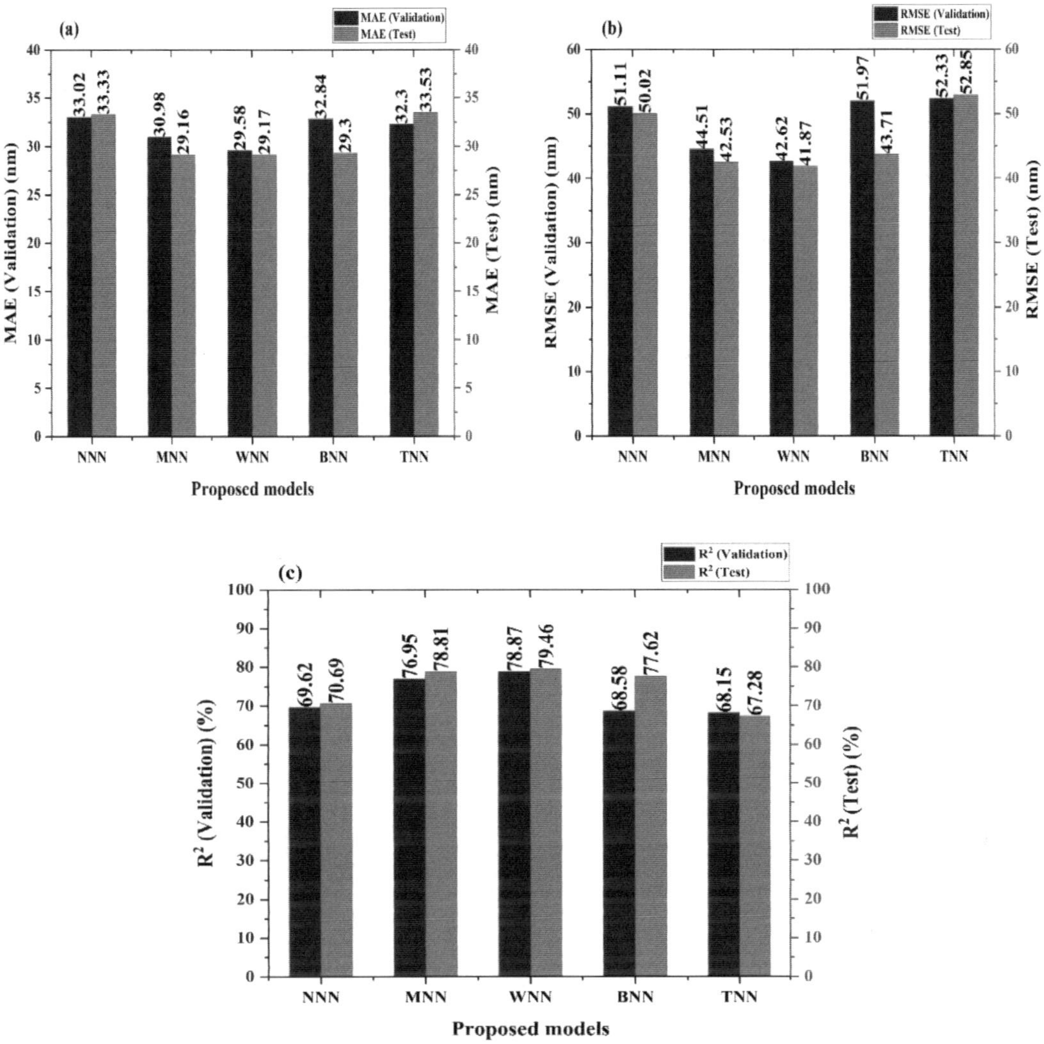

Figure 4. The proposed models' (a) MAE (b) RMSE (c) R^2 of validation and test results for emission maxima prediction.

Table 2. Comparative study of proposed models on (a) MAE, (b) RMSE, and (c) R^2 in both validation and test results for emission maxima prediction.

Sl. No.	Proposed models	MAE (Validation) (nm)	RMSE (Validation) (nm)	R^2 (Validation) (%)	MAE (Test) (nm)	RMSE (Test) (nm)	R^2 (Test) (%)
1	NNN	33.02	51.11	69.62	33.33	50.02	70.69
2	MNN	30.98	44.51	76.95	29.16	42.53	78.81
3	WNN	29.58	42.62	78.87	29.17	41.87	79.46
4	BNN	32.84	51.97	68.58	29.30	43.71	77.62
5	TNN	32.30	52.33	68.15	33.53	52.85	67.28

In the context of emission maxima prediction, while comparing the implemented models, two (MNN and WNN) models achieved above 75% accuracy, and three (NNN, BNN, and TNN) models obtained less than 75% R^2 value as per the validation result. The TNN model had the lowest accuracy of 68.15% among the implemented models. Whereas in terms of test scores, four models (NNN, MNN, WNN, and BNN) got good results above 70% accuracy, and the one remaining model (TNN) got less than 70% accuracy. The WNN outperformed the other models in both validation (train) and test results, and their predicted response versus true response results are graphically illustrated in Figure 5.

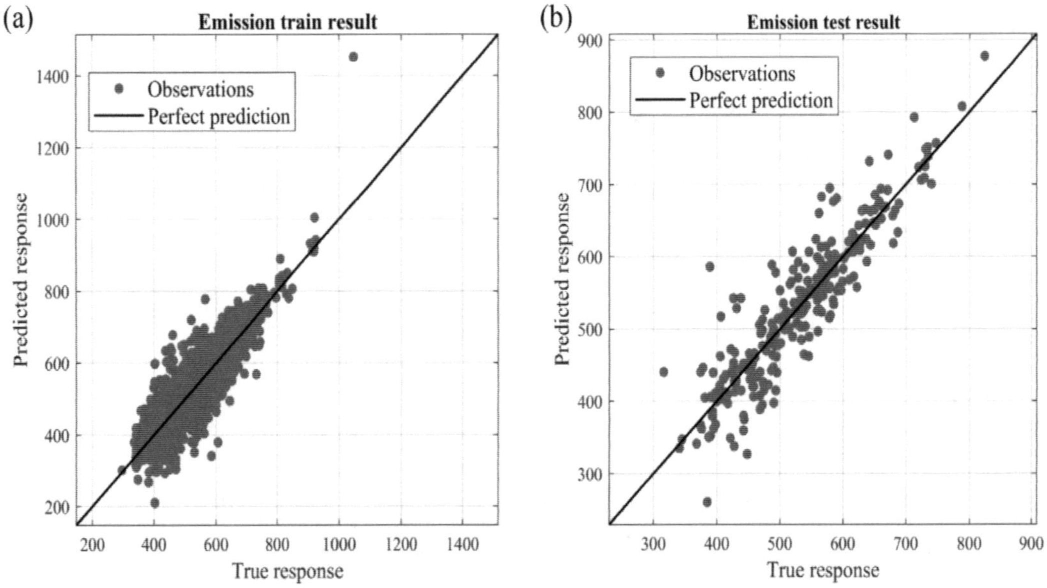

Figure 5. Correlation plots of emission maxima on (a) validation (train) and (b) test results of the WNN model.

The accuracy of NNN, MNN, and WNN gradually increases in both absorption and emission maxima prediction due to stepwise increasing the number of first layer sizes from 10, 25, and 100. Table 3 shows the models' hyperparameter details for each implemented model, along with the layer size of the models. After increasing the number of layer sizes, the model can capture a wider range of patterns. It is capable of a fair mix between model complexity and increasing computing performance. BNN and TNN were facing difficulties interpreting and understanding data due to their connected data layers. It can be harder to discern how the model is working under the hidden layer. Here, in the examined dataset, BNN and TNN results were not good enough in comparison to the other implemented models due to a lack of understanding of the data and a failure to create the proper relation among them after increasing the number of layers 2 or 3.

Table 3. Hyperparameter details of optimized neural network models

Hyperparameter details	Proposed models				
Preset	NNN	MNN	WNN	BNN	TNN
Number of fully connected layers	one	one	one	two	three
layer size (LS)	1st LS: 10	1st LS: 25	1st LS: 100	1st LS: 10 2nd LS: 10	1st LS: 10 2nd LS: 10 3rd LS: 10
Activation	ReLU	ReLU	ReLU	ReLU	ReLU
Iteration limit	1000	1000	1000	1000	1000
Regularization strength (Lambda)	zero	zero	zero	zero	zero
Standardize data	Yes	Yes	Yes	Yes	Yes

This study profoundly studied the absorption and emission maxima of fluorescent organic materials. It is noticeable that the absorption maxima prediction obtained better results than the emission maxima prediction. Because the emission maxima in excited state phenomena are combined with the involvement of radiative and non-radiative processes, which totally depend on the radiative transition rate, the non-radiative transition rate, and other environmental factors. The environmental factors are such as solvent, temperature, reaction temperature, mass of the precursor, ramp rate, reaction time, and so on.

4. Conclusion

Motivated by the design of new fluorescent organic materials with preferred absorption and emission maxima values for several extensive applications such as dye-sensitized solar cells, light harvesting, fluorescence imaging, fluorescence sensing, and so on. Most applications require early and quick prediction of the photophysical properties of materials. In this context, this study considers 3066 fluorescent organic materials to accomplish the objective of absorption and emission maxima prediction. To fulfill the goal, five neural network models (NNN, MNN, WNN, BNN, and TNN) are employed. Tables 1 and 2 tabulate the test results of all implemented models for predicting absorption and emission wavelength maxima. The WNN model is the best predictive model in both validation results and test results of absorption maxima, with R^2 values of 89.55% and 92.14%, respectively. On the other hand, the same model (WNN) is the best predictive model for validation and test results, with accuracies of 78.87% and 79.46%, respectively. These results demonstrate the effectiveness of the suggested optimized neural network models, reducing the need for a trial-and-error methodology in the experimental process. As such, it is an economical method of designing novel fluorescent materials for specific applications. The study's shortcomings include its inability to create new hybrid ensemble models using pre-existing models in the MATLAB Regression Learner App and its rigidity when handling complex datasets.

To enhance the performance of optimized neural network regression models and help scientists create precise medication designs and novel fluorescent organic materials, the current study recommends applying the suggested models to evaluate more fluorescent material datasets in future studies.

Acknowledgements
I (K.D.M.) would like to thank Professor Dr. Raj Nandkeolyar, Department of Mathematics, NIT Jamshedpur, Jharkhand, India, for providing the MATLAB facility. K.D.M. acknowledges the CSIR NET-JRF Fellowship (File No. 09/1277(0001)/2019-EMR-I) for financial support.

Data availability statement
The examined dataset, absorption prediction dataset, and emission prediction dataset are available at https://github.com/KDMsir/Absorption-and-emission-maxima-prediction-using-optimized-neural-network-models

References

[1] Mahato K D and Kumar U 2023 A Review of Organic Dye Based Nanoparticles: Preparation, Properties, and Engineering/Technical Applications *Mini. Rev. Org. Chem.* **20** 655–74
[2] Mahato K D and Kumar U 2023 A comparative study of conventional FRET and light harvesting properties of Rh-110/Rh-6G and Rh-19/Rh-B organic dye pairs impregnated in sol-gel glasses *Methods Appl. Fluoresc.* **11** 035003
[3] Li J, Vacher M, Dral P O and Lopez S A 2023 Machine learning methods in photochemistry and photophysics *Theoretical and Computational Photochemistry: Fundamentals, Methods, Applications and Synergy with Experimental Approaches* (Elsevier) pp 163–89
[4] Tomar N, Rani G, Dhaka V S, Surolia P K, Gupta K, Vocaturo E and Zumpano E 2023 Molecular Structure-Based Prediction of Absorption Maxima of Dyes Using ANN Model *Big*

Data Cogn. Comput. **7** 115
[5] Mai J, Lu T, Xu P, Lian Z, Li M and Lu W 2022 Predicting the maximum absorption wavelength of azo dyes using an interpretable machine learning strategy *Dye. Pigment.* **206** 110647
[6] Shao J, Liu Y, Yan J, Yan Z Y, Wu Y, Ru Z, Liao J Y, Miao X and Qian L 2022 Prediction of Maximum Absorption Wavelength Using Deep Neural Networks *J. Chem. Inf. Model.* **62** 1368–75
[7] Ksenofontov A A, Lukanov M M, Bocharov P S, Berezin M B and Tetko I V. 2022 Deep neural network model for highly accurate prediction of BODIPYs absorption *Spectrochim. Acta - Part A Mol. Biomol. Spectrosc.* **267** 120577
[8] Ju C W, Bai H, Li B and Liu R 2021 Machine Learning Enables Highly Accurate Predictions of Photophysical Properties of Organic Fluorescent Materials: Emission Wavelengths and Quantum Yields *J. Chem. Inf. Model.* **61** 1053–65
[9] Senanayake R D, Yao X, Froehlich C E, Cahill M S, Sheldon T R, McIntire M, Haynes C L and Hernandez R 2022 Machine Learning-Assisted Carbon Dot Synthesis: Prediction of Emission Color and Wavelength *J. Chem. Inf. Model.* **62** 5918–28
[10] Hong Q, Wang X Y, Gao Y T, Lv J, Chen B Bin, Li D W and Qian R C 2022 Customized Carbon Dots with Predictable Optical Properties Synthesized at Room Temperature Guided by Machine Learning *Chem. Mater.* **34** 998–1009
[11] Ye Z R, Huang I S, Chan Y Te, Li Z J, Liao C C, Tsai H R, Hsieh M C, Chang C C and Tsai M K 2020 Predicting the emission wavelength of organic molecules using a combinatorial QSAR and machine learning approach *RSC Adv.* **10** 23834–41
[12] Agarwal T, Mahato K D, Azad C and Kumar U 2022 Predicting Happiness Score During Covid-19 Using Machine Learning *2022 4th International Conference on Artificial Intelligence and Speech Technology (AIST)* (IEEE) pp 1–6
[13] Ju C W, Bai H, Li B and Liu R 2021 Machine Learning Enables Highly Accurate Predictions of Photophysical Properties of Organic Fluorescent Materials: Emission Wavelengths and Quantum Yields *Figshare. Am. Chem. Soc.*
[14] Cereto-Massagué A, Ojeda M J, Valls C, Mulero M, Garcia-Vallvé S and Pujadas G 2015 Molecular fingerprint similarity search in virtual screening *Methods* **71** 58–63
[15] Rogers D and Hahn M 2010 Extended-connectivity fingerprints *J. Chem. Inf. Model.* **50** 742–54
[16] Landrum G 2016 Rdkit: Open-source cheminformatics software
[17] Khan M U, Samer S, Alshehri M D, Baloch N K, Khan H, Hussain F, Kim S W and Zikria Y Bin 2022 Artificial neural network-based cardiovascular disease prediction using spectral features *Comput. Electr. Eng.* **101** 108094
[18] Reichardt C 1994 Solvatochromic dyes as solvent polarity indicators *Chem. Rev.* **94** 2319–58
[19] Catalán J 2009 Toward a generalized treatment of the solvent effect based on four empirical scales: Dipolarity (SdP, a new scale), polarizability (SP), acidity (SA), and basicity (SB) of the medium *J. Phys. Chem. B* **113** 5951–60

IOP Conf. Series: Materials Science and Engineering **1316** (2024) 012012 doi:10.1088/1757-899X/1316/1/012012

Comparison of Hardness by Electrolytic and Electroless Deposition of Nickel on Aluminium

Mrinmoy Chanda, Avik Kar, Sathi Banerjee

Department of Metallurgical and Material Engineering, Jadavpur University, Kolkata-700032, India

corresponding email address: mrinmoy.jumme@gmail.com (M. Chanda)

Abstract: The work deals with the hardness characteristic of substrates coated with nickel through two primary methods: electrolytic deposition and electroless Ni-P deposition. The optimised process parameters for both the process was determined by varying a particular parameter keeping the rest constant in order to obtain minimum crystallite size of Ni. The size of the crystallite were determined using Scherrer's equation. The optimized combination of process parameter for electrolytic deposition is: 270gm/lt of Nickel Sulfate, 60gm/lt Nickel Chloride, 35gm/lt Boric Acid, current density of 90mA/cm^2 and bath temperature of 55°C. The optimized combination of process parameter for electroless Ni-P deposition: 34gm/lt Nickel Sulfate, 23 gm/lt Sodium Hypophosphite, 46 gm/lt of Lactic acid, 42 gm/lt of Sodium Citrate, pH 5.0 and bath temperature 85°C. The optimised deposition rate 22μm/hr. Micro-hardness (Vickers) tests were conducted on the coated samples with a 50 gf load. The cause for difference in hardness was due to buildup of residual stress during deposition of Nickel on the substrate(aluminium). The orientation and surface texture of the materials are determined.

Keywords: Electrolytic and electroless coating, micro-hardness, residual stress, texture analysis, pole figure.

1 Introduction

In recent years, there has been a surge in the development, production, and characterization of nano-crystalline (NC) materials, characterized by crystallites sizes typically smaller than 100 nm. This area of study has garnered significant attention from both scientific and industrial communities. Among the various manufacturing methods, electroplating has emerged as a particularly promising technique due to its feasibility, cost-effectiveness, and economic viability. For over twelve years, electroplating has been employed in the synthesis of bulk NC metals.[1]

However, it has been demonstrated that under specific conditions, only electrodeposition can yield NC nickel coatings. Despite the abundance of studies focused on synthesizing NC nickel deposits, determining the optimum conditions for producing NC nickel coatings remains challenging. Nickel deposition can be achieved through two primary methods: electrolytic deposition and electroless deposition.

Electroless Nickel plating is a process of depositing Nickel-Phosphorus coating over a substrate in the absence of electric current. Coatings are deposited over the substrate by chemical reactions taking place

IOP Conf. Series: Materials Science and Engineering **1316** (2024) 012012 doi:10.1088/1757-899X/1316/1/012012

in the aqueous solution. The dissimilarity between electroless deposition process and the conventional electroplating processes is that the latter depends on an exterior source of dc in order to reduce nickel ions in the electrolyte to nickel metal on the substrate, while the former does not. The engineering industry, encompassing mechanical, electronics, chemical, and printing sectors, is witnessing a surging demand for thicker Electroless Nickel (EN) plating deposits. This heightened demand is driving a quest for enhanced production efficiency, leading to the emergence of super high-speed electroless nickel plating techniques capable of achieving deposition rates of 23-27 μm/h.[4]

EN plating is sought after for its array of benefits including exceptional corrosion resistance, wear resistance, solderability, low porosity, uniform thickness even on intricate parts, and the ability to plate on non-conductive surfaces. Consequently, EN plating finds wide applications across various engineering domains such as oil, chemical, plastics, optics, printing, aerospace, automotive, textiles, shipbuilding, and food industries.

Further residual stress builds up during deposition of Ni on the substrate resulting in change in texture, in plane deformation and change in surface hardness. Presence of residual stress leads to degradation of the coating properties as well as leading to atmospheric corrosion. Thus in order to reduce the residual stress the electroplated sample are annealed to 250°C. This also leads to unifrom distribution of nickel adhesiveness towards the substrate.

The aim of our present work is to provide a consistent set of experimental data within a large range of process parameters for synthesis of NC nickel coatings. Results on the optimization and synthesis conditions of NC nickel deposits from bath are presented. As well as the bath composition has an important effect on the hardness and formation of residual stress on the coated area of the substrate.

This paper distinguishes itself by thoroughly examining the hardness of nickel coatings on aluminum through both electrolytic and electroless deposition methods. Unlike prior studies that typically concentrate on a single technique, this research meticulously optimizes the process parameters for both methods and investigates their influence on crystallite size, hardness, and residual stress. The findings reveal that electroless deposits exhibit higher residual stress due to the formation of a Ni_3P network, which also results in increased hardness compared to electrolytic deposits. Furthermore, the study explores the effects of annealing on residual stress and hardness, providing a detailed analysis that offers new insights into the mechanical properties of nickel-coated aluminium.

2 Experimental Details

2.1 Substrate Preperation

Aluminium Substrate of size 40 x 20 x 0.1 mm^3 is cut from an aluminium foil(99% purity) of grade Al-1198 for the deposition of electroless and electrolytic deposition of Nickel. The aluminum underwent pretreatment, involving a 5-minute immersion in 3.6 M H_2SO_4 at 60°C, followed by rinsing in Sodium hydroxide solution and finally cleaning with acetone and distilled water.[1] The following pre-treatment helps in removal of the aluminium oxide layer formed due to atmospheric oxidation on the surface.

2.2 Bath Preparation

Several factors influence the deposition rate in EN plating, including bath temperature, pH, Ni2+ ion and sodium hypophosphite concentration, bath loading, and the type of complexing agent, accelerator, and stabilizer. Two methods are highlighted for improving the deposition rate: changing the complexing agent to reduce the activation energy needed for Ni2+ complex ion reduction and adjusting the bath composition. Common complexing agents include hydroxyl acids like lactic acid, malic acid, citric acid, and glycolic acid. The Bath composition for electrolytic and electroless Nickel deposition is given in table 1 and table 2[5][4]:

Table 1. Bath composition for Electrolytic Nickel deposition

Nickel sulphate(NiSO4.6H2O)	240-300 gm/lt
Nickel chloride($NiCl_2.6H_2O$)	30-90 gm/lt
Boric acid	30-45 gm/lt
Temperature	40°C - 60°C
pH	3.5-4.5
Cathode current density	2-7 A/dm^2

Table 2. Bath composition for Electroless Nickel deposition

Nickel Sulphate	30-40gm/lt
Sodium hypophosphite	20-28gm/lt
Lactic acid	40-47 gm/lt
Sodium citrate	30-50 ml/lt
pH	5-6.5
Temperature	80-95°C

2.3 XRD analysis

X-Ray diffraction analysis (XRD) is a nondestructive technique that provides detailed information about the crystallographic structure, chemical composition, and physical properties of a material. It is based on the constructive interference of monochromatic X-rays and a crystalline sample. XRD analysis to be carried out to determine the crystallite size of the coating using Scherrer's equation as well to determine the uniformness of the coating. As well as it will carried out to determine phosphorous content during electroless plating of nickel.

The crystallite size of the nickel deposite was determined using the Scherrer's formula. The Scherrer's formula is given as follow:

$$D = \frac{K\lambda}{\beta \cos(\theta)}$$

Where D = crystallite size(nm)
K = Shape factor(0.9)
β = Full Width at half maximum of the peak(FWHM)
λ = X-Ray wavelength(1.54Å)
θ = Diffraction angle(radians)

2.4 Bath compositions for different plating

For electroplating of Nickel bath with varying current density and bath temperature were prepared. The varying conditions are given in the table 3 and table 4. As well the condition for the electroless deposition under varying parameter of the concentration of lactic acid and sodium citrate, pH and bath temperature are given in table 5, table 6, table 7 and table 8.[5]

IOP Conf. Series: Materials Science and Engineering **1316** (2024) 012012 doi:10.1088/1757-899X/1316/1/012012

Table 3. electrolytic deposition with varying current density

Sample index	Nickel sulphate(gm/50 ml)	Nickel Chloride (gm/50ml)	Boric acid (gm/50ml)	Bath temperature	Ph	Current density (A/dm^2)
Sample 1_1	12.5	5.0	2.0	60	3.5	3.2
Sample 1_2	12.5	5.2	2.0	60	3.4	4
Sample 1_3	12.4	5.0	2.0	60	3.5	5
Sample 1_4	12.3	5.0	2.0	60	3.3	6
Sample 1_5	12.5	5.1	2.1	60	3.6	7
Sample 1_6	12.4	5.0	2.0	60	3.5	8

Table 4. electrolytic deposition with varying bath temperature

Sample index	Nickel sulphate (gm/50 ml)	Nickel Chloride (gm/50ml)	Boric acid (gm/50ml)	Bath temperature	Ph	Current density (A/dm^2)
Sample 2_1	12.5	5.0	2.0	45	3.5	7.1
Sample 2_2	12.5	5.2	2.0	50	3.4	7
Sample 2_3	12.4	5.0	2.0	55	3.5	7.1
Sample 2_4	12.3	5.0	2.0	60	3.3	6.8
Sample 2_5	12.5	5.1	2.1	65	3.6	7.2

Table 5. electroless deposition with varying lactic acid concentration

Sample index	Nickel sulphate(gm/ lt)	Sodium Hypophosphite (gm/lt)	Lactic acid(gm/lt)	Sodium Citrate(ml/lt)	Ph	Bath Temperature
Sample 3_1	35	26	42	42	4.9	85
Sample 3_2	35	25.9	44	42.01	5	85
Sample 3_3	35.01	26.03	46	42.03	5.1	85
Sample 3_4	34.9	26	48	41.98	4.9	85

Table 6. electroless deposition with varying sodium citrate concentration

Sample index	Nickel sulphate(gm/ lt)	Sodium Hypophosphite (gm/lt)	Lactic acid(gm/lt)	Sodium Citrate(ml/lt)	Ph	Bath Temperature
Sample 4_1	35	26	46.02	30	4.9	85
Sample 4_2	35	25.9	45.89	35.01	5	85
Sample 4_3	35.01	26.03	45.92	39.98	5.1	85
Sample 4_4	34.9	26	46.05	45.05	4.9	85
Sample 4_5	35	25.9	45.98	50.01	5.0	85

Table 7. electroless deposition with varying pH

ample index	Nickel sulphate(gm/ lt)	Sodium Hypophosphite (gm/lt)	Lactic acid(gm/lt)	Sodium Citrate(ml/lt)	Ph	Bath Temperature
Sample 3_1	35	26	46	42	3.5	85
Sample 3_2	35	25.9	46	42.01	4.5	85
Sample 3_3	35.01	26.03	45.98	42.03	5	85
Sample 3_4	34.9	26	45.93	41.98	5.5	85
Sample 5_5	35.02	26	46.01	42.08	6.0	85
Sample 5_6	34.98	25.98	46.02	41.99	6.5	85

Table 8. electroless deposition with varying Bath temperature

Sample index	Nickel sulphate(gm/lt)	Sodium Hypophosphite (gm/lt)	Lactic acid(gm/lt)	Sodium Citrate(ml/lt)	Ph	Bath Temperature
Sample 6_1	35	26	46	42	5.0	75
Sample 6_2	35	25.9	46	42.01	4.9	80
Sample 6_3	35.01	26.03	45.98	42.03	5.0	85
Sample 6_4	34.9	26	45.93	41.98	4.9	90
Sample 6_5	35.02	26	46.01	42.08	5.1	95

2.5 Texture analysis

During the deposition process of Ni, columnar crystallitess commonly form and then induce a fibrous surface texture. Because of the anisotropy of the columnar crystallites, the crystallites sizes are believed to be different along different directions, implying that relationships exist between the deposition parameters and texture. Thus the residual stress and crystallite misorientation was determined in the following experiment.

3 Result & Discussion

3.1 XRD Analysis

The XRD data of the different samples undergoing electrolytic and electroless deposition with varying parameter are given below. The crystallite size is found using Scherrer's formula and crystallite size is plotted with varying parameter. As well as crystallite size was plotted against differing plating parameter.

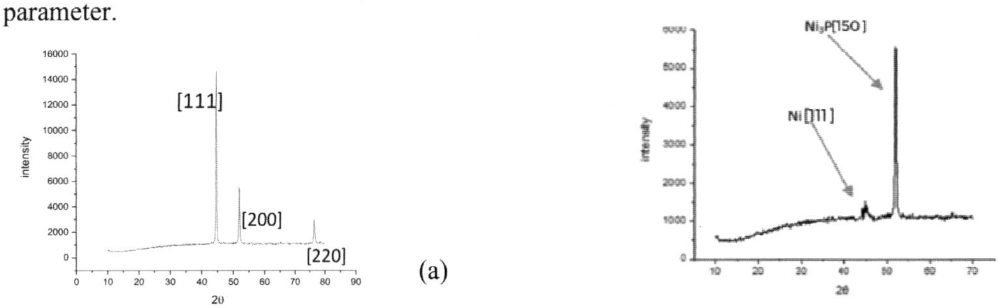

Fig 1: (A) XRD of electrolytic deposition of nickel, (B) XRD of electroless deposition of nickel

Fig1(A) reveals the Ni peaks of electrolytic deposition of nickel. The XRD data reveals presence of Ni with (111) orientation at 44.5° (JCPDs card no: 03-1051), (200) orientation at 51.9°(JCPDs card no: 04-0850) and (220) orientation at 76.4°(JCPDs card no: 88-2326). Fig1(B) reveals the Ni_3P and Ni peaks due to electroless nickel deposition. The Ni peaks were suppressed by the formation of Ni_3P peak. So the only visible peak of Ni is at 44.5°. XRD data reveals presence of Ni with (111) orientation at 44.5° (JCPDs card no: 03-1051) and Ni_3P with (150) orientation at 51.93° (JCPDs card no: 89-2743).

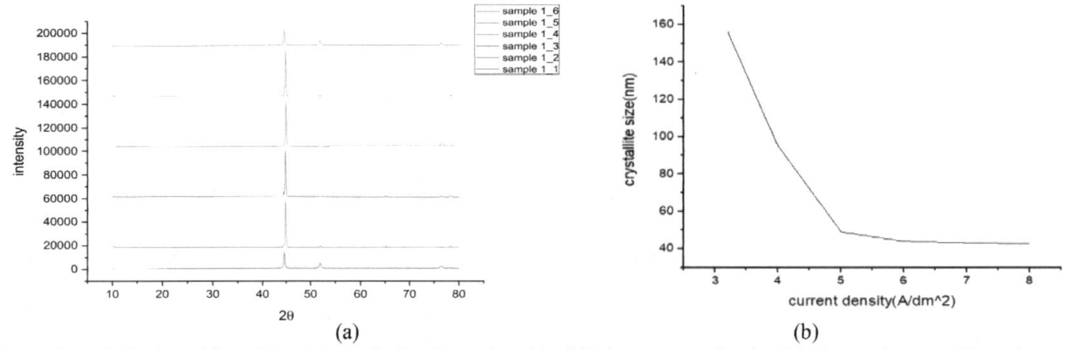

Fig 2:Electrolytic deposition: (a) XRD analysis of sample with differing current density,(b) change in crystallites size with different current density

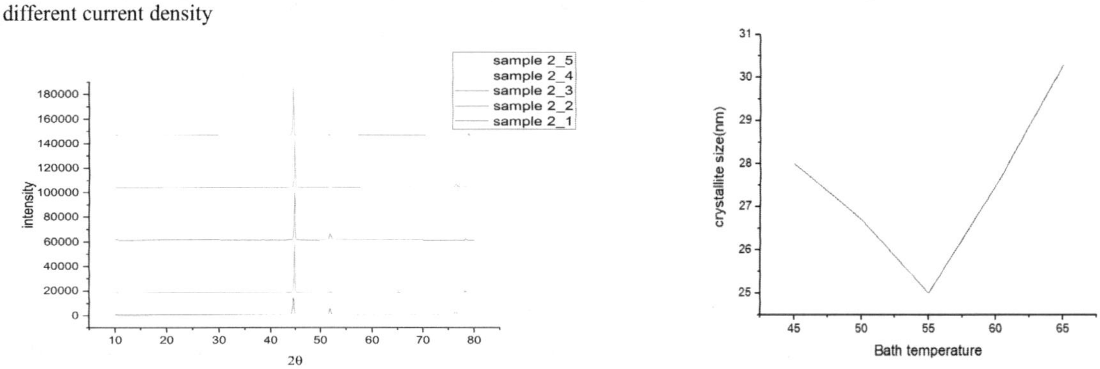

Fig 3: Electrolytic deposition:(a) XRD analysis of sample with differing bath temperature, (b) Change in crystallites size with differing bath temperature

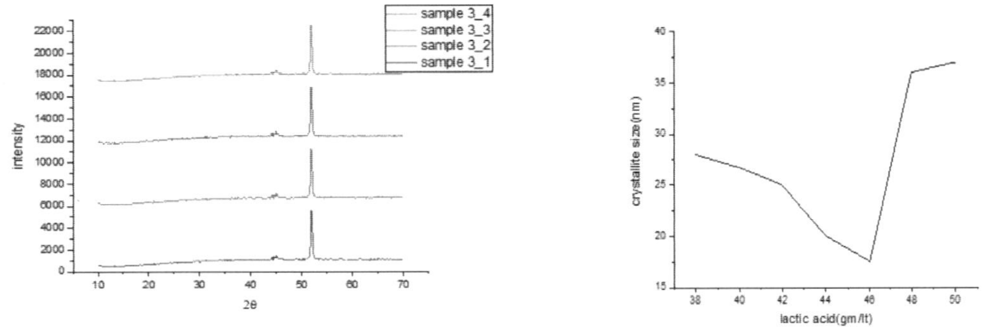

Fig 4: Electroless deposition: (a) XRD analysis of sample with varying lactic acid conc, (b) Change in crystallites size with varying lactic acid concentration

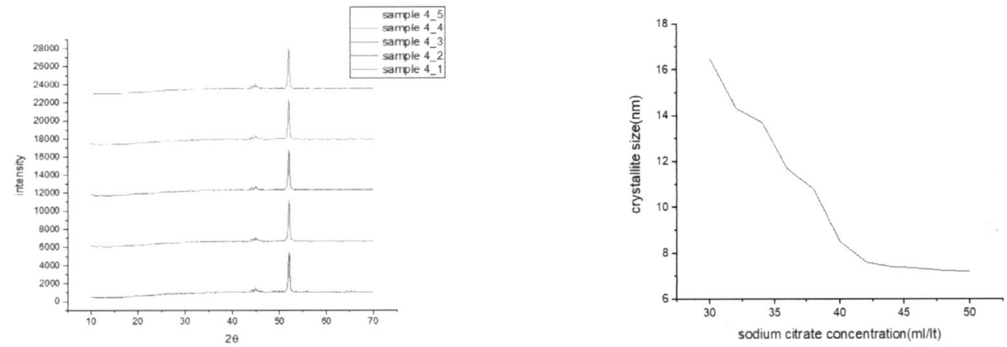

Fig 5: Electroless deposition: (a) XRD analysis of sample with varying sodium citrate conc, (b) Change in crystallites size with varying sodium citrate concentration

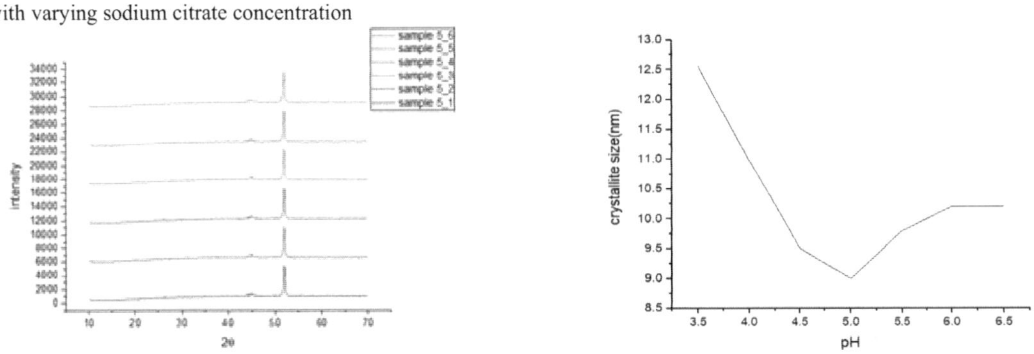

Fig 6: Electroless deposition: (a) XRD analysis of sample with varying pH, (b) Change in crystallites size with varying pH

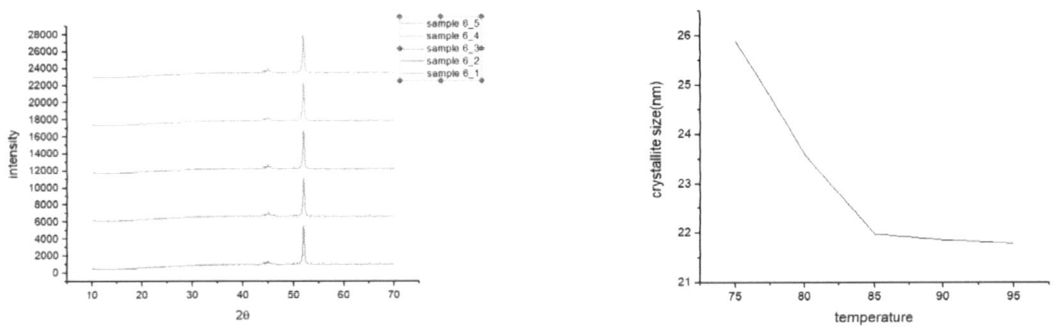

Fig 7: Electroless deposition: (a) XRD analysis of sample with varying Temperature, (b) Change in crystallites size with varying Temperature

Fig 2 and fig 3 shows stacked plot of the XRD(Fig 2-A & Fig 3-A) obtained by changing only the current density and bath temperature keeping other factors for electrolytic nickel deposition almost same. The grain size were estimated using Scherrer's equation for the largest peak at 44.50°. Fig 2-B shows that the grain refines when the current density is increased to 5A/dm^2, but further increase shows no significant change in grain size. The grain refines with increasing current density is probably due to increase to higher overpotential further increasing the deposition rate of the nickel crystallites. Fig 3-B shows the variation of crystallite size with change in bath temperature. It is observed that the grain size decreases upto a temp of 55°C and then increases with a stiff rise. Increasing the bath temperature has two contradictory effect: (1) an increase in critical size due to decrease in thermodynamic driving force of the crystallization process which leads to decrease in nucleus density, and (2) an increase in kinetic energy that can lead to increase of deposition rate and nucleation rate. Due to this below 55°C the deposition was governed by thermodynamic factor whereas above it the deposition was governed by kinetic theory. Thus for electrolytic deposition we find that with increasing current density above 5A/dm^2 the crystallite size remains almost same. With bath temperature minimum crystallite size is obtained at 55°C. Thus the optimum condition for electrolytic nickel deposition is: 250gm/lt Nickel Sulphate, 60 gm/lt nickel chloride, 40gm/lt boric acid, pH 4.0-4.3, bath temperature 55°C and cathodic current density of 5A/dm^2.

For electroless nickel Ni-P deposition we find that first the crystallite decreases upto 46 gm/lt of lactic acid and then further increase exponentially which maybe due to decrease in pH and formation of carboxylic deposition along with Ni-P. when concentration of sodium citrate increases from 42 ml/lt the crystallite size become almost unchangeable. Minimum crystallite size is found at pH 5 keeping all other processing parameters constant. Again when bath temperature increases from 85°C the change in crystallite size become almost unchangeable. Thus the optimum process parameter in order to form smallest crystallite size is: 35gm/lt nickel sulphate, 26gm/lt sodium hypophosphite, 46 gm/lt lactic acid, 42 ml/lt sodium citrate, pH equals to 5 and bath temperature of 85°C

3.2 Micro-Vicker Hardness: Hardness of the Ni coated substrate was measured by vicker hardness test using a load of 5gf force. The hardness(GPa) versus the variation of the process parameter is plotted in the following graphs.[7]

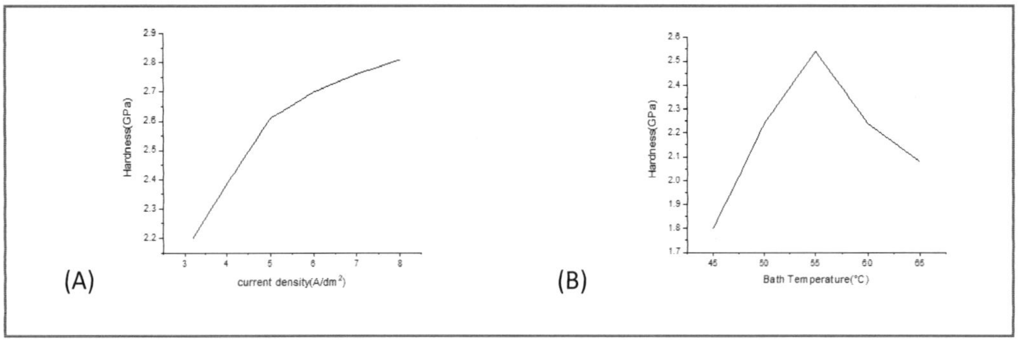

Fig 8. The hardness of nc-Ni coatings as a function of the variation of the process parameter: (A) Change in current density and (B) change in bath temperature

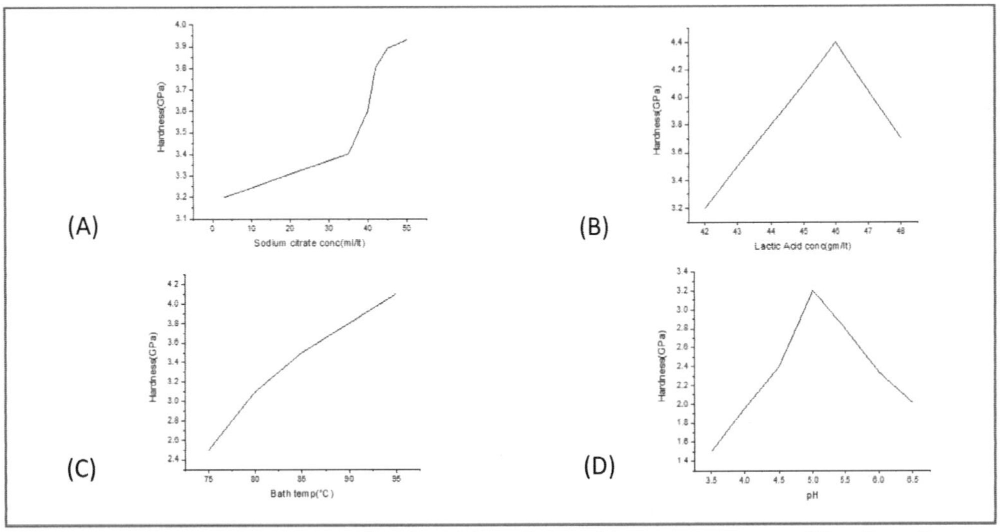

Fig 9: The hardness of Ni-P coatings as a function of the variation of the process parameter: (A) sodium citrate concentration, (B) Lactic acid concentration, (C) Bath temperature and (D) pH value

Fig 8 shows the variation of hardness of the nc-Ni coating with varying process parameter. Vicker hardness was converted to GPa and plotted against the process parameter. The hardness initially increases exponentially with increase in current density upto 5A/dm^2, above which the hardness didn't increased to much extent and remains almost same(Fig 8-A). The maximum hardness achieved due to varying current density is 2.7 GPa. Fig 8-B shows the variation of hardness with bath temperature for electrolytic nickel deposition. It is observed that at around 55°C maximum hardness is achieved. This is because at 55°C the crystallite size are minimum. The maximum hardness achieved is approx. 3.55 GPa.

3.3 Surface texture analysis: Hardness of the electrodeposited substance increase primarily due to formation of residual stress[6]. But development of excess residual stress lead to degradation of the electrodeposited because it may result in reduction of adhesiveness between the substrate and the electrodeposited material. Surface texture of the nickel deposited Al substrate were analysed in order to determine the residual stress distribution due to optimized electrodeposition of Nickel on Al substrate. The stress distribution is given as follow. As well the orientation of crystallite along [111], [200] and [220] crystallographic direction is given by its pole figure. The pole figure of material(substrate+nickel deposited) was taken for as-received and after annealing at 240°C.

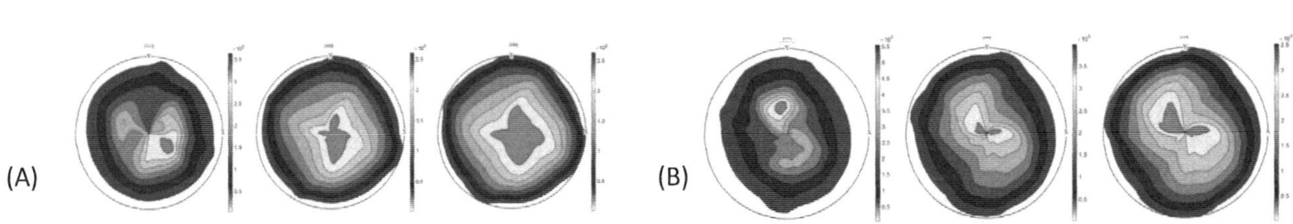

Fig 10. Pole Figure distribution nc-Ni crystallite distribution due to optimized electrolytic deposition: (A) As-Received and (B) annealed at 240°C

IOP Conf. Series: Materials Science and Engineering **1316** (2024) 012012 doi:10.1088/1757-899X/1316/1/012012

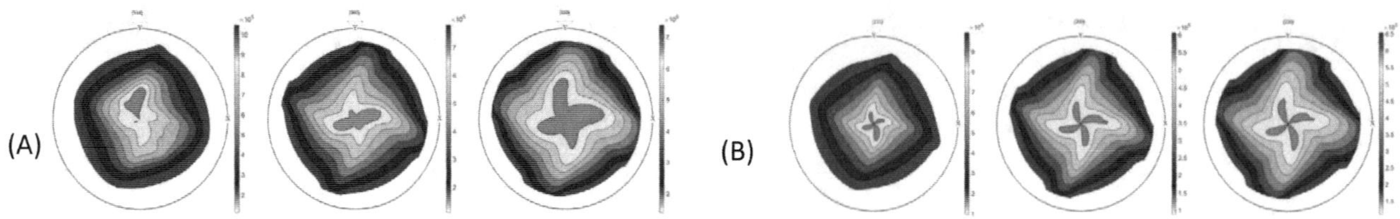

Fig 11. Pole Figure distribution Ni-P crystallite distribution due to optimized electroless deposition: (A) As-Received and (B) annealed at 240°C

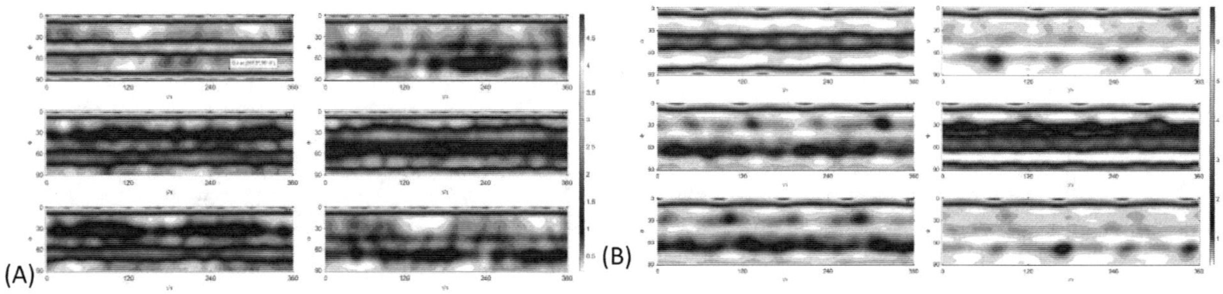

Fig 12.ODF of nc-Ni distribution on the substrate due to optimised electrolytic deposition: (A) As-Received and (B) annealed at 240°C

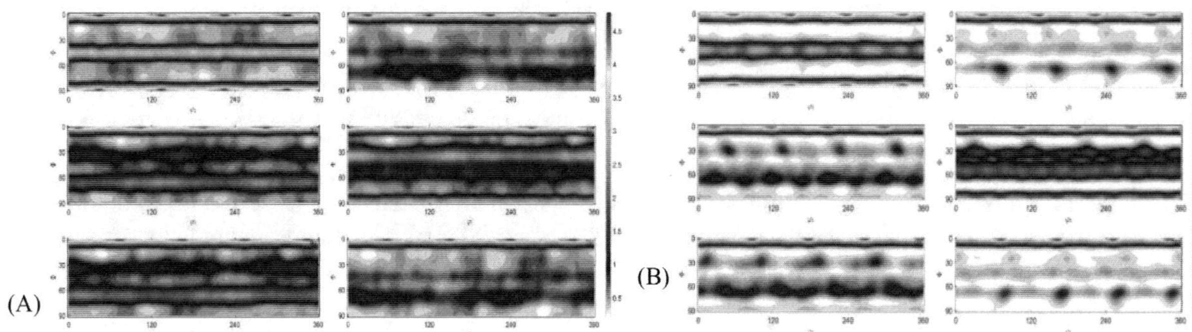

Fig 13. ODF of Ni-P distribution on the substrate due to optimised electroless Nickel deposition: (A) As-Received and (B)annealed at 240°

From Fig-10(A) we found that the crystallite distribution of Ni was not uniform throughout the substrate and due to which variation of hardness may be observed at different location of the substrate. This gives rise to degradation of the plating resulting in easy removal of the nickel plating from the substrate due to non-uniformity in thickness. This may also result in rough surface morphology resulting a critical site for crevice and pitting corrosion. When annealed at 240°C(fig 10-B) the crystallites in order to reduce to increase adhesiveness rearranged themselves resulting in uniformity in crystallite directions and thickness and uniform distribution. The pole figure of the annealed electrodeposited material reveals that the crystallites are distributed uniformly with intensity maximum at the centres and decreasing toward the corners. Fig 12 shows the distribution of residual stress for electrodeposited as-received and annealed material at different crystallographic direction along x-z plane. As-received electrodeposited(fig 12-A) shows high residual stress due to high cathode current density formation on the substrate. When annealed at 240°C the stress was relieved. Thus after electrodeposition of Ni annealing is necessary otherwise the desirable properties of the electrodeposited materials wouldn't be achieved and would result in easy aerial corrosion and removal of the deposition.

From fig-11 it is observed that crystallites having crystallographic direction of [220] and [200] are more or less distributed uniformly. But the crystallite with direction [111] shows non uniform arrangement which is evident due to non-symmetry of the pole figure. This would result with retardation of surface properties with time[6]. When annealed at 240°C(fig 11-B) the symmetry in crystallite distribution is gained resulting in improvement of surface morphology. Similar to electrodeposition of nickel, in case of electroless deposition the residual stress formed due to deposition(fig 13-A), resulting in degradation of surface morphology and resulting in easy corrosion. The residual stress is relieved by annealeing at 240°C.

But electroless deposited material shows higher residual stress as compared to electrodeposited nickel deposited material. The reason behind it lies due to formation of crystalline Ni_3P and Ni deposition due to electroless nickel deposition, whereas only Ni is deposited in electrolytic deposition of Ni. Ni_3P forms strong continuous network which increases hardness as well as surface stress. But on annealing the residual stress for electrolytic deposited material is more compared to electroless deposited material. This is due to breaking of continuous network of Ni_3P and formation of discrete Ni_3P nickel deposition. The difference in residual stress can be seen by analysing fig 12 and fig 13. The residual stress formed due to electrolytic deposition is 795±49 MPa and on annealing the residual stress comes down to 616±40 MPa. The residual stress of As-received electroless deposited material in 917±30 MPa which comes down to 493±30 MPa on annealing at 240°C. The decrease in residual stress in probably due to transformation of amorphous Ni_3P into crystalline Ni_3P.

4 Conclusion

The optimised paramters for electrolytic and electroless nickel deposition were determined. Hardness was measured by micro-vicker hardness and the hardness change was observed with the inverse of the square root of the crystallite size. One of the interesting factor regarding the difference in hardness between the two type of deposition which is often ignored is enclosed and discussed a function of residual stress formation. Along with the heat treatment to be incorporated to desire the properties for surface deposition was found out. The main conclusions can be enclosed in the following points:

- The criteria for determining the optimised condition for electrolytic and electroless deposition of Nickel on aluminium substrate was to find minimum crystallite size. The optimised condition for electrolytic nickel deposition is: 250gm/lt Nickel Sulphate, 60 gm/lt nickel chloride, 40gm/lt boric acid, pH 4.0-4.3, bath temperature 55°C and cathodic current density of $5A/dm^2$. The optimised condition for electroless deposition of nickel is: 35gm/lt nickel sulphate, 26gm/lt sodium hypophosphite, 46 gm/lt lactic acid, 42 ml/lt sodium citrate, pH equals to 5 and bath temperature of 85°C.
- The hardness variation of was taken by micro-vicker hardness and was plotted against crystallite size. It was seen that with decreasing crystallite size the hardness was increasing hence satisfying Hall-Petch relation. Additionally the hardness observed for electroless nickel deposition was more than electrolytic nickel deposition due to formation of continuous network of Ni_3P. The hardness comparison reverses when both the material are annealed to a temperature of 240°C.
- Higher degree of crystallite misorientation was observed for electrolytic than that of electroless nickel deposition. Thus indicating electroless nickel deposition results in better surface morphology than that of electrolytic nickel plating.

- As-received electroless deposited material had higher residual stress of about 917MPa, while the material deposited via electrolytic nickel deposition had residual stress of only 795MPa. But on annealing heat treatment the residual stress of electroless deposited material was relieved to a large extent due to breaking of the Ni_3P continuous network. Thus as a result for desirable nickel deposition, electroless deposition should incorporated with above stated optimum parameters further processed by desirable heat treatment schedule.

5 Acknowledgement

The authors would like to thank Jadavpur University for giving us the equipments and the required chemicals for our work. I would further like to extend my thanks to Mr. Avik Kar for helping in carrying out the experiments. Also I would like to thank Dr. Sathi Banerjee for supervision of the experimental works and revised manuscript.

6 References

1. Mechanism of electrodeposition of nickel(https://hal.science/hal-03637891)
2. The Heat Treatment effect of Nickel-coated Steel Plate Substrate by Electroless Plating on Hardness Test(DOI:: 10.14419/ijet.v7i4.33.23559)
3. Analysis of the Characteristics of Nickel-Plating Baths (http://www.ccsenet.org/ijc)
4. Electroless nickel plating from acid bath(DOI: 10.1016/S0257-8972(99)00423-5)
5. The properties and the effect of operating parameters on nickel plating (review)(DOI:10.5897/IJPS11.1163)
6. Direct Correlations among the Grain Size, Texture, and Indentation Behavior of Nanocrystalline Nickel Coatings(https://doi.org/10.3390/met9020188)
7. Metal plating processes and methods of measuring surface hardness and thickness of coating
8. Wang, L.; Gao, Y.; Xu, T.; Xue, Q. A comparative study on the tribological behavior of nanocrystalline nickel and cobalt coatings correlated with grain size and phase structure.
9. Sribalaji, M.; Asiq Rahman, O.S.; Laha, T.; Keshri, A.K. Nanoindentation and nanoscratch behavior of electroless deposited nickel-phosphorous coating.
10. Bund, A.; Thiemig, D. Influence of bath composition and pH on the electrocodeposition of alumina nanoparticles and nickel
11. 11. Godon, A.; Creus, J.; Cohendoz, S.; Conforto, E.; Feaugas, X.; Girault, P.; Savall, C. Effects of grain orientation on the Hall–Petch relationship in electrodeposited nickel with nanocrystalline grains
12. Haciismailoglu, M.; Alper, M. Effect of electrolyte pH and Cu concentration on microstructure of electrodeposited Ni–Cu alloy films. Surf. Coat. Technol. 2011, 206, 1430–1438
13. Rashidi, A.M.; Amadeh, A. Effect of electroplating parameters on microstructure of nanocrystalline nickel coatings. J. Mater. Sci. Technol. 2010, 26, 82–86
14. Watson, S. A. (1992). Electroless nickel coatings. Technical Series No. 10055, Nickel Develop Institute
15. Taheri, R. (2003). Evaluation of electroless nickel – phosphorus (EN) coatings. PhD thesis, University of Saskatchewan

Enhancing Anti-Blast Capabilities of Pipelines with High-Performance Composite CFRP Laminate: Computational Fluid Dynamics-Based Predictive Analysis

Farman Saifi[1], Mohd Javaid[1], Abid Haleem[1] and S. M. Anas[2]

[1] Department of Mechanical Engineering, Jamia Millia Islamia (A Central University), New Delhi, 110025, India
[2] Department of Civil Engineering, Jamia Millia Islamia, New Delhi, 110025, India

E-mail: farmansaifi96@gmail.com

Abstract. Recent explosions highlighted the need for a thorough investigation to understand the impact of these explosions on critical infrastructural systems and networks like pipelines. This paper provides an essential understanding for developing safety protocols and interventions, strengthening resilience, enhancing public safety, mitigating socio-economic consequences, and ensuring sustained progress. Shallow buried conduits, crucial for transporting water, oil, and gas over long distances, face explosion risks due to conflicts, sabotage, terrorism, corrosion, and accidental damage. Recent blast incidents in San Francisco and Lagos underscore the devastating effects of conduit malfunctions, necessitating rigorous safety measures and technological innovations. In this research, an empty pipeline measuring 26.20 mm in thickness and subjected to a detonation initiated upon its uppermost surface is emulated, utilising the sophisticated computational capabilities of Abaqus software, integrating principles of computational fluid dynamics. Following a meticulous validation process achieved via mesh refinement analysis, the study expands to scrutinise two additional detonation scenarios — positioned above and below the pipe's surface — while adhering to a steadfast standoff distance of 25mm. The primary objective of this scholarly endeavour is to ameliorate the deleterious repercussions inflicted upon the conduit by explosive forces, envisioning a scenario where deformation occurs without immediate rupture. To this end, a 0.15 mm thick and 200mm wide laminate composed of composite CFRP with 0°/90° fibre orientation is strategically affixed onto the upper half segment of each pipe model at the impacted region, serving as a fortification measure aimed at bolstering areas of susceptibility and potentially diminishing the magnitude of ensuing blast-induced impairment. Comparative analysis of the pipe's performance, both with and without the applied laminate, is meticulously conducted to discern the effectiveness of this intervention. The results demonstrate a significant reduction in plastic damage with the utilisation of laminate, with models integrating CFRP experiencing an average decrease of approximately 39.94% compared to those without it. It highlights the efficacy of CFRP in protecting pipeline structural integrity against blast events.

1. Introduction

In recent years, there has been a notable escalation in the occurrence of explosions, characterised by abrupt and vigorous release of energy [1-3]. This concerning trend presents significant challenges to the resilience of infrastructure systems. The upsurge in explosive incidents underscores the urgency for a comprehensive inquiry into their causative factors, mechanisms, and consequences, particularly from the perspective of mechanical and civil engineering [1-4]. Comprehending the ramifications of explosions on critical infrastructure, including subterranean pipelines, is paramount in this domain. Through meticulous analysis of these ramifications, it becomes feasible to devise sophisticated safety protocols and engineering interventions aimed at averting catastrophic damage and preserving the integrity of essential infrastructure networks. This proactive stance serves to bolster infrastructure resilience, bolster public safety, and mitigate the socio-economic fallout from such occurrences. A thorough investigation into the effects of explosions is indispensable for fortifying infrastructure against unforeseen perils and ensuring the sustainability of societal progress.

Buried shallow conduits represent integral facets of our infrastructure, serving as indispensable conduits for conveying crucial resources, including water, oil, and natural gas [5-6]. These intricate networks constitute the vital framework underpinning diverse industries, facilitating the expedient dispersion of raw materials over extensive distances. However, notwithstanding their indispensability, shallowly buried conduits are susceptible to detonations owing to many factors, such as corrosion, inadvertent infliction of damage, or even deliberate acts of subversion [5-6]. The ramifications of conduit impairment or rupture are profound and far-reaching. Such occurrences disrupt the seamless flow of resources and pose significant perils to adjacent communities and ecosystems. Recent incidents, such as the 2019 explosion in a residential enclave triggered by a ruptured gas conduit in San Francisco, California, and the 2022 pipeline detonation in Lagos, Nigeria, underscore the catastrophic aftermaths of conduit malfunctions. In both instances, these explosions resulted in loss of lives, property devastation, and extensive disruption to infrastructure networks. To mitigate such hazards and fortify the anti-explosive resilience of conduits, stringent safety protocols and technological innovations are imperative. Implementing measures such as advanced conduit monitoring systems, materials resistant to corrosion, and bolstered emergency response frameworks can curtail the probability of explosions and mitigate their deleterious impacts on infrastructure networks and public well-being.

Testing buried pipelines amid explosions is challenging due to resource constraints and safety hazards [2, 7-8]. Academic institutions often need more suitable facilities, and the high costs of specialised testing environments limit research. However, advancements in computer simulations offer a promising alternative. Simulations provide valuable insights into safety measures and risk mitigation strategies by accurately modelling pipeline behaviour. Thus, while experimental tests face constraints, computer simulations offer a viable means of studying buried pipeline dynamics in explosions.

The burgeoning utilisation of Carbon Fiber Reinforced Polymer (CFRP) to augment the blast resilience of diverse infrastructural systems underscores its pivotal significance in mitigating the deleterious effects of explosions. Historical investigations [9-12] consistently underscore the exceptional mechanical attributes of CFRP, including its notable strength-to-mass ratio and exceptional fatigue endurance, rendering it an optimal choice for fortifying structures against blast forces [12]. The expansive corpus of literature on CFRP utilisation vividly illustrates its efficacy in reinforcing the blast resistance of edifices, spans, and other indispensable infrastructure elements. In the current research, this versatile material assumes a central role as CFRP laminate is employed in fortifying pipelines to enhance their resistance to blasts. Through scrupulous scrutiny, this study endeavours to quantify the degree to which CFRP laminate bolsters the pipeline's resilience against blasts, assessing both the mitigation of inflicted damage and the enhancement of structural response. This investigation represents a pivotal stride in fortifying critical infrastructure against the omnipresent menace of explosions, proffering insights into innovative approaches for augmenting blast resilience via advanced material solutions such as CFRP.

Within infrastructure safety considerations, simulating explosion effects on crucial systems such as pipelines is imperative for hazard comprehension and mitigation. Computational fluid dynamics (CFD), notably the Eulerian-Lagrangian method [6, 8, 13] coupled with the Finite Element Method (CEL-FEM) in software like Abaqus [13], offers distinct advantages over empirical methods like the ConWep blast program. Unlike empirical approaches relying on simplified assumptions, CEL-FEM provides a more accurate portrayal of complex fluid-structure interactions during explosions. This methodology combines the Eulerian domain for fluid behaviour tracking with the Lagrangian domain for monitoring particle or structure motion within the fluid [13]. By integrating these domains, CEL-FEM captures blast wave dynamics interacting with pipelines, offering detailed insights into pressure distributions, structural responses, and potential failure modes. This research involves validating CEL-FEM in Abaqus against existing literature results to enhance safety protocols and infrastructure resilience by predicting real-world scenarios reliably.

2. Literature

A significant paucity of scholarly investigation has been evident in underground infrastructure, particularly pipelines vulnerable to explosions. This deficiency contrasts with the escalating frequency of accidental detonations, intentional acts of sabotage like VBIED attacks and the deployment of suitcase bombs, all unfolding against a backdrop of global conflicts. The necessity for a thorough understanding, predictive capabilities for response, and effective mitigation strategies regarding these systems has become increasingly conspicuous. Designers and researchers are awakening to the critical importance of addressing these vulnerabilities, precipitating a shift towards developing robust methodologies to safeguard underground infrastructure from diverse explosive threats. With these threats continually evolving, an ever more pressing need arises for innovative solutions and interdisciplinary collaboration in this domain. Below is a succinct overview of previous research endeavours on pipelines conducted by various researchers.

Song et al. [6] examined X70 grade steel pipelines' response to localized blast loading, identifying four failure modes influenced by explosive mass, contact area, and wall thickness. Increased mass and contact area led to higher deflection and damage, while wall thickness affected post-failure motion. Numerical simulations corroborated experimental findings. Xue et al. [14] proposed a theoretical model predicting deep-buried pipeline displacement from blast-induced ground shock. Factors like soil-pipeline interaction and ground shock distribution were investigated, with LS-DYNA simulations confirming model reliability. Wu et al. [15] evaluated ground particle and pipe dynamics through explosion tests and 3D numerical models. Control standards based on dynamic stress yield criteria were established to prevent pipeline damage, emphasizing safety measures near rock-soil interfaces during blasting. Qin et al. [16] assessed soil and pipeline responses to explosions using vibration monitoring and simulations. Findings informed a predictive model and safety criterion formula for future safety assessments during construction. Jiang et al. [17] analyzed blasting excavation for Wuhan Metro Line 8 stage two, proposing a mathematical model for peak particle velocity attenuation with depth. A predictive model for peak von Mises stress under blast vibrations was established to ensure pipeline safety during nearby construction. Zhao et al. [18] conducted blast experiments on a buried gas pipeline in an urban area, favoring interfaceless pipelines for their lower vulnerability to blast vibrations. Wu et al. [19] investigated gas steel pipeline burial in Chinese cities, determining criteria to prevent failure from surface explosive charges.

The current study endeavours to fill a notable void in extant scholarly research by directing attention towards enhancing the anti-blast prowess of subterranean compromised pipelines through the utilisation of CFRP laminate—a hitherto unexplored methodology in antecedent investigations. While antecedent scholarly endeavours have thoroughly scrutinised various facets of pipeline reaction to detonations, encompassing dynamic characteristics, subsoil particle kinetics, soil-pipeline interplay, and safety evaluations, none have specifically delved into the employment of CFRP laminate to augment the anti-blast capabilities of the pipelines. This innovative approach heralds a prospective resolution for ameliorating blast-induced impairments to buried pipelines, proffering strides in pipeline safety and fortitude against explosions.

3. Establishing the Modeling and Simulation Framework

The numerical framework crafted within the Abaqus [13] platform integrates diverse components pivotal for replicating explosive incidents entailing a steel conduit. These components comprise an unoccupied steel conduit/pipeline with a thickness measuring 26.20mm, a brick of TNT explosive, a soil mass supporting the conduit, an Eulerian space encompassing air and TNT, and a Lagrangian mesh delineating both the subterranean mass and the conduit, Figure 1. In this configuration, the Eulerian domain, excluding its basal facet, implements non-reflective boundary conditions (NRBC) to obviate wave reflections at its periphery, facilitating the faithful emulation of an unbounded environment. Concurrently, the basal face is rigidly constrained across all degrees of freedom, mimicking the characteristics of rigid ground, thereby safeguarding stability and furnishing a robust basis for the simulation.

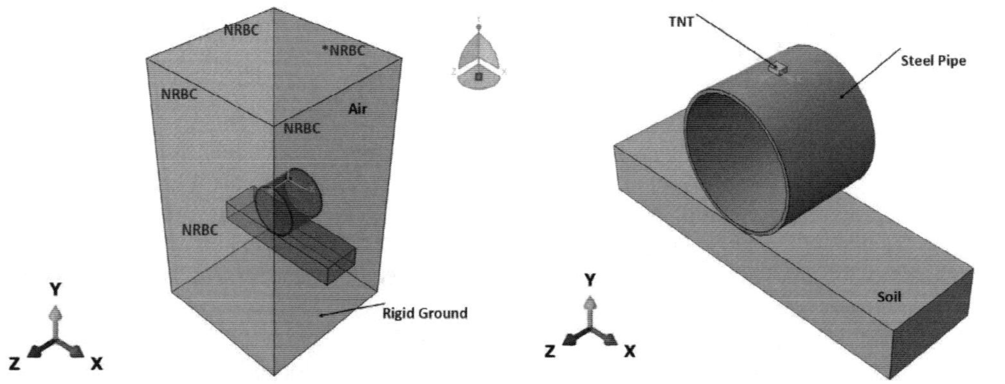

Figure 1. Simulation framework.

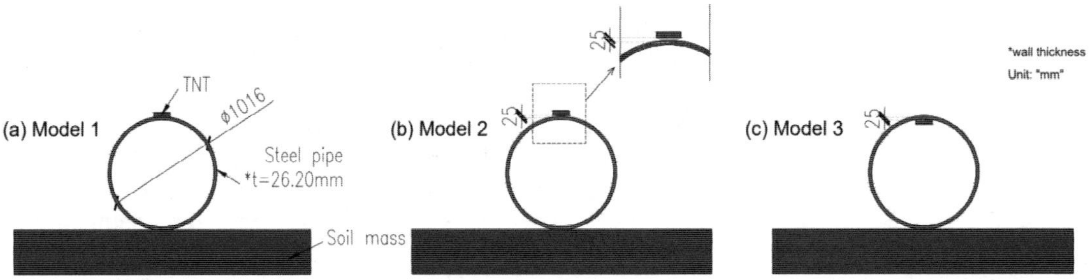

Figure 2. Pipeline models without CFRP retrofit.

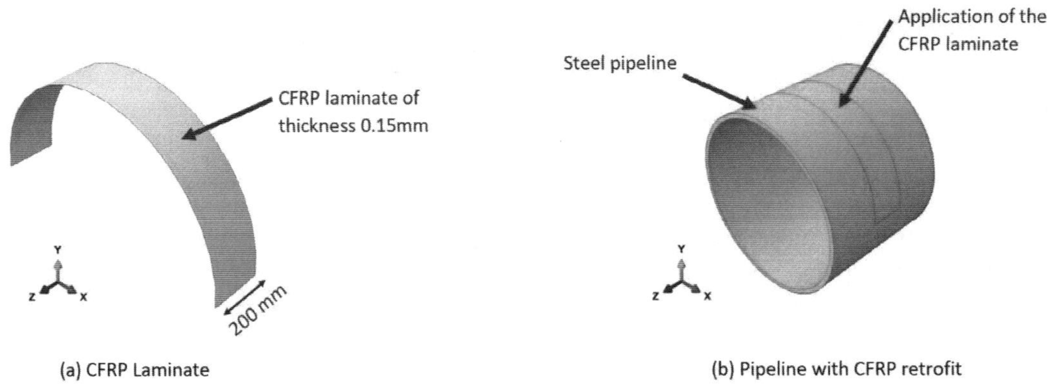

Figure 3. Pipeline model with laminate of CFRP.

The standard contact-to-contact interface between the subterranean mass and the conduit is adopted, streamlining the simulation of their interaction during the explosive occurrence. This interaction mechanism facilitates a precise portrayal of the kinetic forces and distortions manifesting at the juncture between the conduit and the adjacent soil. The modelling approach meticulously mirrors the experimental configuration delineated in a specified reference [6], ensuring concordance with established empirical findings. Placing the TNT explosive in direct contact with the upper top surface of the conduit at its midpoint emulates the experimental parameters stipulated in reference [6], thereby preserving coherence and comparability between the simulation and experimental outcomes. Following meticulous analyses involving mesh refinement and validation against empirical data, the investigation encompasses two supplementary blast scenarios beyond the primary contact detonation. These scenarios entail detonation of the TNT explosive above and below the upper top surface of the conduit at a blast distance of 25mm, Figure 2. This separation distance emulates close-range detonations typical in industrial or urban settings, thereby furnishing insights into conduit integrity's immediate and formidable repercussions under such circumstances. In response to the possibility of pipeline distortion during extreme loading, namely contact, above-surface, and below-surface explosions, three distinct models depicting damage are generated within Abaqus software. Anticipating deformation without immediate rupture, efforts are directed towards alleviating blast-induced effects on the pipeline. To this end, a CFRP laminate [9-12], boasting a thickness of 0.15mm, has been meticulously applied to each model. This strategic intervention aims to fortify vulnerable regions, potentially diminishing the magnitude of subsequent blast-related damage. Consequently, six models are scrutinised: three devoid of CFRP reinforcement, juxtaposed with the remaining three fortified by CFRP enhancement. This method facilitates a comparative examination of CFRP's efficacy in abating blast ramifications across various explosive scenarios, furnishing invaluable insights into potential strategies for preserving pipeline integrity in environments susceptible to detonations.

Below are the six models considered in this research, each designated accordingly to reflect variations in explosive scenarios and the presence or absence of CFRP reinforcement (Figure 3).
Model 1: Primary contact detonation scenario without CFRP (Control/Reference Model); *Model 2*: Above-surface detonation without CFRP; *Model 3*: Below-surface detonation without CFRP; *Model 4*: CFRP reinforcement (i.e., in the form of lamination) on *Model 1*; *Model 5*: CFRP reinforcement on *Model 2*; *Model 6*: CFRP reinforcement on *Model 3*.

The use of CFRP reinforcement to ameliorate the impact of explosions on pipelines bears notable pragmatic significance in real-world contexts. Strengthening compromised segments of pipelines with CFRP presents industries with the prospect of bolstering infrastructural network durability and curtailing operational interruptions resulting from blast incidents. This approach augments safety protocols and fosters economically efficient maintenance and risk mitigation measures within pivotal infrastructure domains like the oil and gas industry, thus guaranteeing uninterrupted functionality even in environments prone to explosions.

In this study, the constituents within the simulation are delineated by specific constitutive models within the framework of Abaqus. Steel is delineated utilising the Johnson-Cook plasticity methodology, which comprehensively addresses the material's robustness and deformation rate dependencies, particularly under conditions of substantial strain. The Mohr-Coulomb plasticity model encapsulates the soil's behaviour, which adeptly captures its response to varying stress levels and delineates failure criteria. On the other hand, air is symbolised by the Ideal Gas Equation of State (EoS), a fundamental relationship elucidating the interplay among pressure, volume, and temperature for compressible fluids, such as air. Lastly, TNT is explicated via the Jones-Wilkins-Lee EoS, a model tailored to accommodate its markedly energetic and compressible properties, thereby facilitating a precise simulation of its explosive characteristics. The simulation leverages material characteristics from diverse authoritative sources to ensure precision and dependability. In particular, the mechanical attributes of steel are meticulously extracted from reference [6], guaranteeing a faithful portrayal of its structural response. The soil parameters, critical for precise modelling, are gleaned from reference [13, 20], offering a thorough compendium of soil mechanics data. Material properties pertinent to TNT and air are sourced from references relevant [6, 8, 13], thus cementing the simulation's adherence to real-world conditions. Such scrupulous curation of material data from reputable sources bolsters the veracity and relevance of the simulation outcomes within practical contexts. The steel utilised in the simulation possesses a strength of 560 MPa, a density of 7850 kg/m^3, and an elastic modulus of 210 GPa. Additionally, a dynamic increase coefficient 1.24 is applied to account for strain rate effects, aligning with blast design standards outlined in UFC 3-340-02 [21] for a more accurate representation of steel behaviour under dynamic loading conditions.

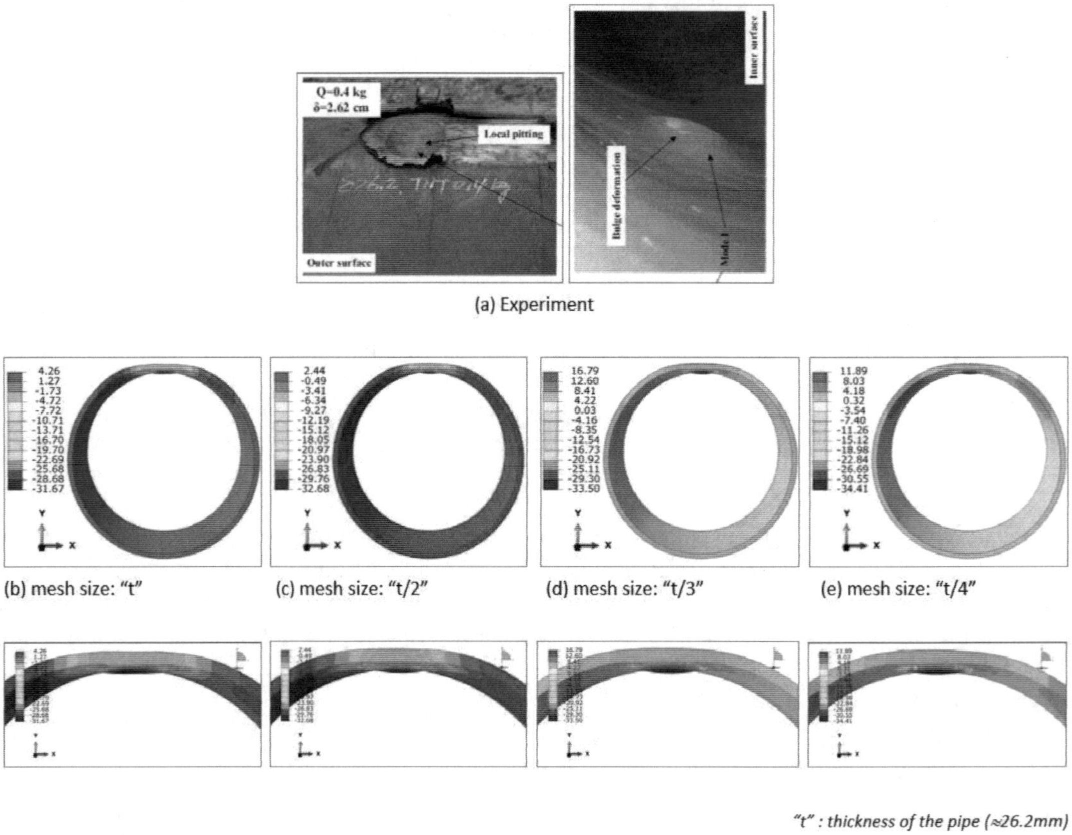

Figure 4. Validation of numerical pipe deformation (mm) of Model 1.

In simulating CFRP laminate behaviour with Hashin's damage [13] criteria in Abaqus, diverse failure modes like fibre fracture, matrix cracking, and fibre-matrix debonding are comprehensively considered. This model evaluates failure based on stress states and material properties. Using S4R elements, tailored explicitly for thin shell structures like laminates, efficiently captures composite behaviour while maintaining computational efficiency. Representing cohesive interaction between the pipe and laminate with a tie constraint in Abaqus ensures structural continuity without separation. Material properties of the laminate are sourced meticulously [10], enhancing simulation accuracy and consistency. The laminate exhibits a remarkable tensile strength of 2200 MPa alongside a mass density of 1800 kg/m^3, as indicated by reference [10, 22].

The dimension of finite elements holds significant sway in finite element analysis, especially concerning model authentication. Optimal element sizing is paramount to ensuring precision and efficacy in depicting the performance of the simulated structure or substance. In this investigation, the C3D8R element variant represents soil, pipe, TNT, and air, presenting a harmonious blend of computational expediency and accuracy. Furthermore, the EC3D8R element type is enlisted for the Eulerian mesh, furnishing a resilient portrayal of fluid dynamics and interactive phenomena. This meticulous curation of element types augments the faithfulness and dependability of the simulation outcomes, thereby facilitating the efficient validation of the model against experimental observations.

The ensuing segment substantiates the reference model by implementing mesh refinement analysis (MRA). Various dimensions of finite elements for the conduit/pipe denoted as "*t*", "*t/2*", "*t/3*", and "*t/4*" (where "*t*" signifies the thickness of the conduit) are scrutinised. The resultant distortions stemming

IOP Conf. Series: Materials Science and Engineering **1316** (2024) 012013 doi:10.1088/1757-899X/1316/1/012013

from these varying element dimensions are juxtaposed against extant data to gauge the precision and dependability of the model. Below are the categorisations of mesh dimensions employed in the analysis, designated thus: "*Standard*" aligns with the pipe's thickness "*t*"; "*Moderate*" with half of the pipe's thickness "*t/2*"; "*Fine*" with one-third of the pipe's thickness "*t/3*"; and "*Very Fine*" with one-fourth of the pipe's thickness "*t/4*".

4. Analysis of Mesh Refinement

MRA has been conducted in this section, as depicted in Figure 4. The pipe model has undergone a TNT contact blast with a mass of 0.40 kg, aligning with the investigation conducted by Song et al. [6]. Notably, the maximal deformation escalated as the dimension of elements diminished, as discerned. The "*Very Fine*" mesh exhibited a maximal deformation of 34.41 mm, 1.68% inferior to the blast test data. Conversely, the "*Standard*" mesh manifested a deformation of 31.67 mm, constituting a 9.51% deviation from the experimental data [6]. Moreover, the "*Fine*" and "*Moderate*" mesh types yielded deformations of 33.50 mm (4.28% lower than the test data) and 32.68 mm (6.63% lower than the test data), respectively. Given its minimal deformation deviation (1.68%), the "*Very Fine*" mesh was deemed suitable for scrutinising the pipe's response to explosions, thus being employed for the parametric analysis expounded in this paper. The rise in maximum deformation with smaller mesh sizes is due to increased resolution, capturing localised deformations and stress concentrations more accurately. However, the disparity between simulated and experimental data arises from the trade-off. While smaller meshes offer better accuracy, they require more computational resources, leading to longer execution times and increased storage needs.

In Figure 4, the computer simulation reveals slight discrepancies between expected and actual results due to various factors, notably idealised boundary conditions and omitting weather effects like wind and temperature. Limited experimental data on pipelines under blast loading necessitated reliance on referenced studies for comparison.

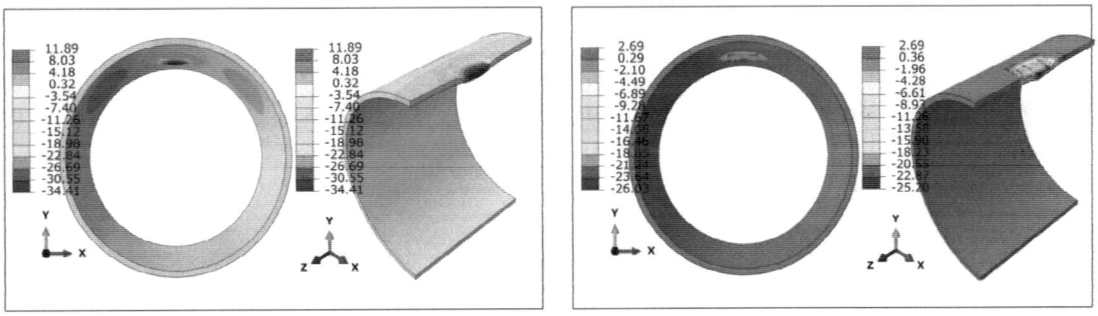

(a) Model 1 (b) Model 4

Figure 5. Pipe deformations (mm) (a) without and (b) with CFRP: contact blast.

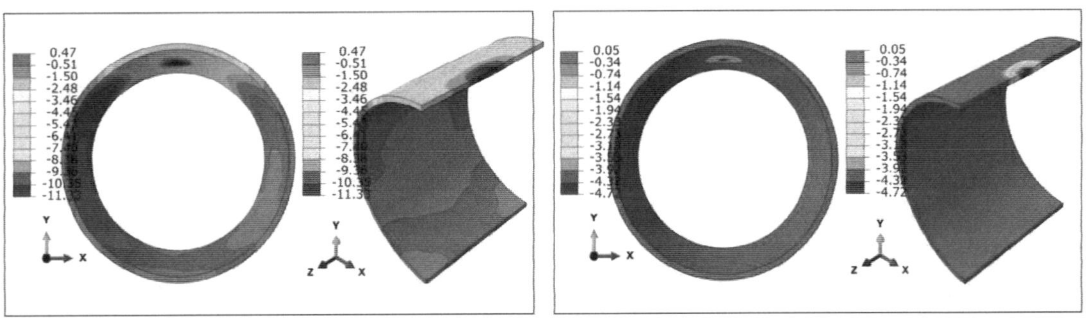

(a) Model 2 (b) Model 5

Figure 6. Pipe deformations (mm) (a) without and (b) with CFRP: above-surface blast.

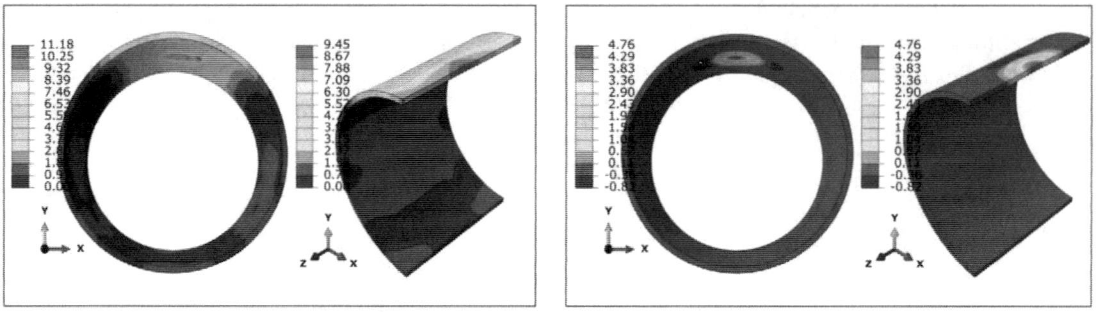

(a) Model 3 (b) Model 6

Figure 7. Pipe deformations (mm) (a) without and (b) with CFRP: below-surface blast.

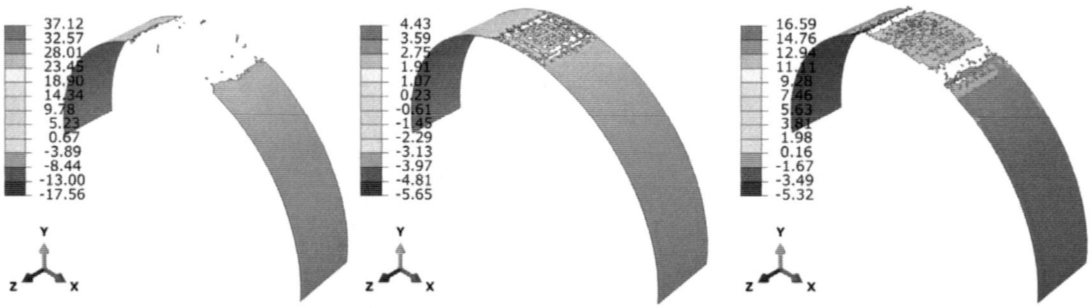

(a) Model 4 (b) Model 5 (c) Model 6

Figure 8. Deformations (mm) and rupturing of CFRP.

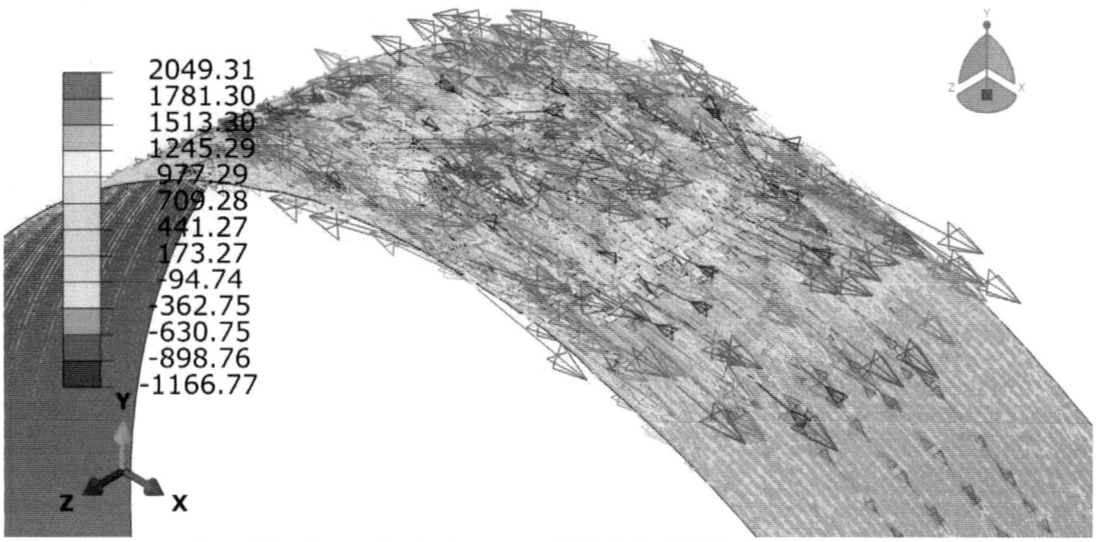

Figure 9. Formation of in-plane principal stresses (MPa) in CFRP for dissipating explosion energy: Model 6 is taken as an example in this figure.

5. Results

When subjected to a contact blast at its upper top surface, the empty pipe undergoes a complex series of dynamic responses. Initially, the detonation of the explosive generates a compression wave that propagates through the air and interacts with the surface of the pipe. This compression wave causes a rapid increase in pressure on the pipe's surface, leading to deformation. As the compression wave travels through the pipe material, it generates high tensile stress waves within the pipe structure. These high tensile stress waves create internal forces that result in significant deformation of the pipe. In this scenario, the pipe experiences a deformation of 34.41mm (Figure 5), indicating the magnitude of structural displacement due to the blast impact. The plastic damage incurred by the pipe, quantified as 119.51 Joules, reflects the energy absorbed by the material during the deformation process. This plastic damage value signifies the extent of permanent deformation and structural alteration experienced by the pipe due to the contact blast. In Model 3, the maximum upward deformation measures 9.45mm, whereas in Model 2, due to varying blast scenarios, there is a downward deformation of 11.33mm. Deformations in the pipeline without and with CFRP are shown in Figure 5, Figure 6, and Figure 7. Models 4, 5, and 6 experience peak deformations of 26.03mm (downward), 4.72mm (downward), and 4.76mm (upward), respectively. These values are 24.35%, 58.34%, and 49.62% lower, respectively, than those observed in their corresponding models without CFRP. Figure 8 clearly demonstrates that the laminate acts as a sacrificial layer for the pipeline, substantially reducing explosion energy and deformation. The development of in-plane stresses within the laminate assists in dispersing explosion energy, consequently reducing the amount of energy transferred to the pipeline. This leads to an enhanced pipeline response, as illustrated in Figure 9.

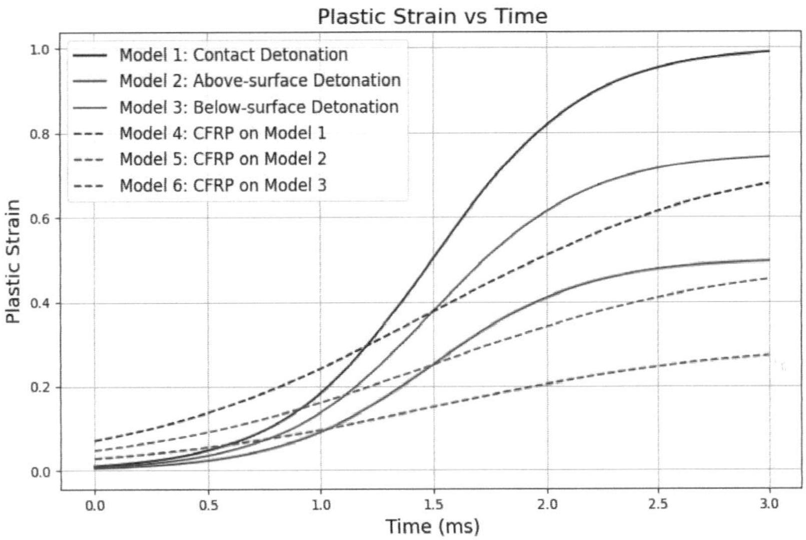

Figure 10. Plastic strain in the pipe material is at the impacted region.

The enhancement in the pipeline's anti-blast response upon reinforcement with CFRP laminate is attributable to several pivotal factors:

- **Energy Dissipation**: Laminate's inherent strength and stiffness properties enable it to absorb a considerable amount of energy during explosive events. Acting as a buffer, the laminate dissipates a substantial portion of the explosion's force through in-plane action (Figure 9), thereby reducing the energy transferred to the pipeline.
- **Strengthening Effect**: The application of laminate fortifies the pipeline's integrity, effectively bolstering its resistance against blast loads. As an additional protective layer, the laminate facilitates a more uniform distribution of applied forces across the pipeline's surface, mitigating displacement and plastic damage by offering supplementary support to the pipeline structure.
- **Load Redistribution**: laminate facilitates a more efficient redistribution of applied loads throughout the pipeline structure. By reinforcing the upper half of the pipeline, the laminate redirects blast forces away from critical areas, such as vulnerable points, diminishing the risk of failure or rupture.

- **Diminished Displacement**: By absorbing and dissipating a significant portion of explosion energy, laminate reduces the overall displacement of the pipeline. Consequently, the pipeline experiences diminished deformation and movement in response to blasts, resulting in reduced structural damage and deformation.
- **Mitigation of Plastic Damage**: The incorporation of laminate aids in mitigating plastic damage within the pipeline material. By absorbing energy and dispersing loads more uniformly, the laminate diminishes the likelihood of localised plastic deformation or failure, thereby preserving the pipeline's structural integrity.
- **Augmented Resilience**: In essence, the application of laminate augments the pipeline's resilience against blast loads. By effectively dissipating energy and reinforcing the structure, the laminate minimises the risk of catastrophic failure, enhancing the pipeline's survivability and longevity in high-risk environments.

Upon reinforcing a pipeline with laminate, the detonation of an explosion instigates the propagation of compression waves throughout the structure. Upon encountering the laminate, characterised by its formidable tensile strength and stiffness attributes, it resists deformation and adeptly absorbs a substantial proportion of the explosive energy. This laminate efficiently disperses the applied forces, thwarting their concentration within specific pipeline segments. Furthermore, the exceptional tensile strength inherent in CFRP staunchly opposes the tensile stresses provoked by the explosion, thereby fortifying the pipeline's structural integrity. The interplay between the compression waves and the laminate yields reflected waves, which dissipate energy within the laminate further and preclude the transmission of deleterious forces back into the pipeline. Consequently, this process effectively mitigates displacement and plastic damage, augmenting the pipeline's resilience against blast loads while preserving its structural integrity.

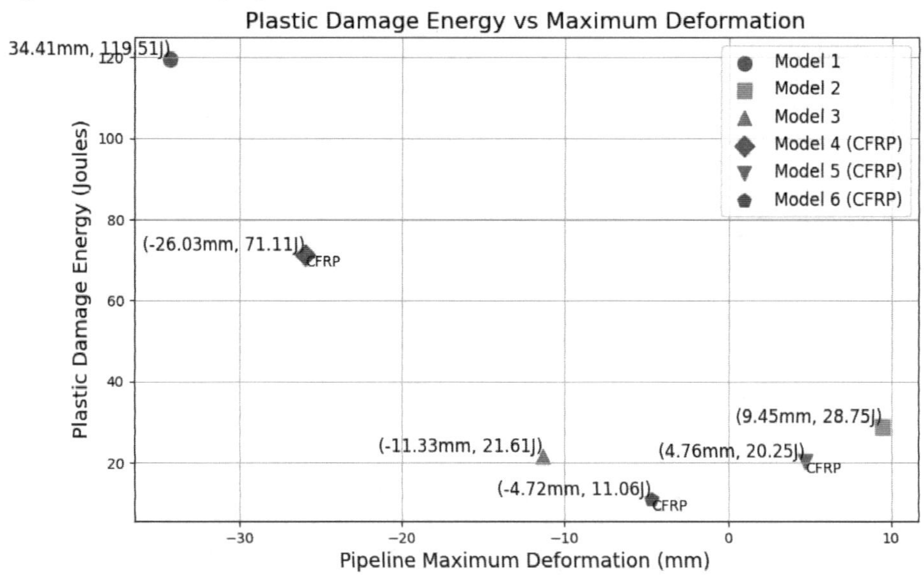

Figure 11. Plastic damage versus pipe deformation.

The laminate, when applied on a pipeline, mitigates plastic strain experienced during a blast event, Figure 10. This reduction in plastic strain enhances the pipeline's ability to withstand explosive forces without enduring permanent deformation or damage. It is imperative to acknowledge that in the event of a contact explosion, the laminate is wholly ruptured or its properties are compromised, contrasting with the lesser severity of rupture observed in the other two blast events. This discrepancy arises from the thickness of the laminate, which can be augmented to mitigate such occurrences.

Figure 11 illustrates the correlation between plastic damage and pipe deformation across six designated models. Models 1, 2, and 3, devoid of laminate application, exhibit plastic damage values of 119.51J, 28.75J, and 21.61J respectively. Contrastingly, the integration of laminate onto the pipes results in a notable reduction in plastic damage, with values of 71.11J, 20.25J, and 11.06J observed in Models 4, 5, and 6, respectively. The mean decrease in plastic degradation across models featuring CFRP laminate in contrast to those devoid of such composite material stands at an approximate magnitude of 39.94%.

6. Conclusions

The pipeline's heightened resistance to explosions after the utilisation of CFRP laminate arises from its adeptness in absorbing energy, redistributing loads, diminishing displacement, alleviating plastic deformation, and fortifying overall resilience against explosive forces. The utilisation of laminate yields a notable mitigation of plastic damage, as illustrated by an average decrease of roughly 39.94% among models integrating CFRP in contrast to those devoid of it. This underscores the effectiveness of CFRP in safeguarding the structural integrity of pipelines in the face of blast occurrences. The mechanism behind CFRP laminate involves the development of in-plane principal stresses within the laminate, which efficiently dissipate explosion energy as compression waves from the blast encounter the CFRP; they deform, generating in-plane stresses that evenly distribute the power across the material's surface. This process effectively reduces the energy transferred to the pipeline structure, mitigating displacement and plastic damage while enhancing resilience against blast loads. While it is evident that during the analysed blast occurrences, the laminate is diminishing its characteristics due to heightened explosive force, this concern can be remedied by augmenting the thickness of the laminate. Moreover, further investigation can be conducted to elucidate supplementary methodologies for enhancing the efficacy of CFRP laminate in attenuating blast impacts on pipelines. Furthermore, it is recommended that the laminate be used when pipelines endure damage without rupture during explosive events. This utilisation entails fortifying compromised segments of pipelines to mitigate further degradation and augment their resilience against subsequent blasts.

Limitations: The limitation of this investigation is evident in its exclusive focus on utilizing CFRP laminate to enhance pipeline resistance to explosions. It neglects to explore alternative methods or materials for similar outcomes, nor does it address potential degradation of CFRP characteristics under heightened explosive forces. Furthermore, the recommendation to use CFRP laminate solely for fortifying compromised pipeline segments overlooks potential proactive applications.

References

[1] Anas, S. M., Alam, M., and Umair, M. (2021). "Experimental and Numerical Investigations on Performance of Reinforced Concrete Slabs under Explosive-induced Air-blast Loading: A state-of-the-art review". *Structures*, Elsevier, 31, pp. 428-461, DOI: 10.1016/j.istruc.2021.01.102.

[2] Hao, H., Hao, Y., Li, J., and Chen, W. (2016). "Review of the current practices in blast-resistant analysis and design of concrete structures". *Advances in Structural Engineering*, SAGE, 19(8): 1193-1223.

[3] Anas, S. M., Al-Dala'ien, R. N., Akram, S., and Alam, M. (2024). "Accidental Explosions, Beirut ANFO Blast, Ghana Truck Explosion, Global Terrorism Index 2021-22, and Blast Mitigation: A Review". *5th International Conference on Energetics, Civil and Agricultural Engineering (ICECAE 2024)*, Tashkent, Uzbekistan, May 13- 14, 2024, Vol. 497, Publisher: E3S Web Conf., DOI: 10.1051/e3sconf/202449702001.

[4] Anas, S. M., Alam, M., and Umair, M. (2021). "Air-blast and ground shockwave parameters, shallow underground blasting, on the ground and buried shallow underground blast-resistant shelters: A review". *International Journal of Protective Structures*, SAGE, 13(1), pp. 99-139, DOI: 10.1177/2F20414196211048910.

[5] Jiang, N., Zhu, B., Zhou, C., Li, H., Wu, B., Yao, Y., and Wu, T. (2021). "Blasting vibration

effect on the buried pipeline: A brief overview". Engineering Failure Analysis, vol. 129.

[6] Song, K., Long, Y., Ji, C. (2016). "Experimental and numerical studies on the deformation and tearing of X70 pipelines subjected to localised blast loading". *Thin-Walled Struct.* 107, pp. 156–168.

[7] Kyei, C., and Braimah, A. (2017). "Effects of transverse reinforcement spacing on the response of reinforced concrete columns subjected to blast loading". *Engineering Structures*, 142, pp. 148-164.

[8] Anas, S. M., Shariq, M., Alam, M., and Umair, M. (2022). "Evaluation of Critical Damage Location of Contact Blast on Conventionally Reinforced One-way Square Concrete Slab applying CEL-FEM Blast Modeling Technique". *International Journal of Protective Structures*, SAGE, 13(4), pp. 672-715.

[9] Tahzeeb, R., Alam, M., Anas, S. M., and Muddassir, S. M. (2023). "Dynamic Response of CFST Column with In-plane Cross-reinforcement and Partial CFRP Wrapping upon Contact Blast". *Innovative Infrastructure Solutions*, Springer. DOI: 10.1007/s41062-023-01201-x.

[10] Kadhim, M. M. A., Jawdhari, A. R., Altaee, M. J. et al. (2020). "Finite element modelling and parametric analysis of FRP strengthened RC beams under impact load". *Journal of Building Engineering*, vol. 32.

[11] Haderi, M., Maleki, V. A., and Andalibi, K. (2015). "Retrofitting of unreinforced masonry walls under blast loading by FRP and spray on polyurea". *Cumhuriyet Science Journal*, 36(4).

[12] Anas, S. M., and Alam, M. (2022). "Performance of brick-filled reinforced concrete composite wall strengthened with C-FRP laminate(s) under blast loading". *Materials Today: Proceedings*, Elsevier, 65(1), pp. 1-11, DOI: 10.1016/j.matpr.2022.03.162.

[13] Abaqus CAE (2020). "Damage Plasticity, explicit platform, material library, interactions, constraints, boundary conditions, loads, post-processing". Help & Learning manual, ABAQUS DS-SIMULIA.

[14] Xue, H., Chen, W., Ma, J., Xie, T., and Xu, Z. (2023). "A theoretical modelling study on deep buried pipelines subjected to blast-induced ground shock". *Arabian Journal of Geosciences*, vol. 16.

[15] Wu, T., Jiang, N., Zhou, C., Luo, X., Li, H., and Xia, Y. (2021). "Dynamic response and safety assessment of buried gas pipe subjected to ground surface explosion16". *International Journal of Pressure Vessels and Piping,* 194(Part A).

[16] Qin, T., Wu, M., Jia, L., Xie, L., and Wu, L. (2023). "Effects of Vibration on Adjacent Pipelines under Blasting Excavation". *Applied Sciences*, MDPI, 13(1).

[17] Jiang, N., Zhu, B., He, X., Zhou, C., Luo, X., and Wu, T. (2020). "Safety assessment of buried pressurised gas pipelines subject to blasting vibrations induced by metro foundation pit excavation". *Tunnelling and Underground Space Technology*, vol. 102.

[18] Zhao, K., Jiang, N., Zhou, C., Li, H., Cai, Z., and Zhu, B. (2022). "Dynamic behavior and failure of buried gas pipeline considering the pipe connection form subjected to blasting seismic waves". *Thin-Walled Structures*, vol. 170.

[19] Wu, T., Jiang, N., Zhou, C., Luo, X., and Sun, J. (2021). "Evaluate of anti-explosion for high-pressure gas steel pipeline subjected to ground explosion". Journal of Constructional Steel Research, vol. 177.

[20] Mokhtari M. and Nia A. V. (2016). "The application of CFRP to strengthen buried steel pipelines against subsurface explosion", *Soil Dynamics and Earthquake Engineering*, 87: 52-62.

[21] UFC 3-340-02 (2008), "Structures to resist the effects of accidental explosions", Technical Manual, Unified Facilities Criteria UFC 3-340-02, U.S. Army Corporations of Engineers.

[22] Anas, S. M., Alam, M., and Umair, M. (2022). "Strengthening of Braced Unreinforced Brick Masonry Wall with (i) C-FRP Wrapping, and (ii) Steel Angle-Strip System under Blast Loading". Materials Today: Proceedings, Elsevier, 58(4), pp. 1181-1198, DOI: 10.1016/j.matpr.2022.01.335.

IOP Conf. Series: Materials Science and Engineering **1316** (2024) 012014 doi:10.1088/1757-899X/1316/1/012014

Material Selection and Optimization of Green Drilling of Aluminium Alloys: MCDM-Pareto ANOVA and T-SRC

KMB. Karthikeyan[1*], Sam Richard N[2], Bharath S[3], Cyril Moses L[4], Eshwar NK[5]

[1, 2, 3, 4, 5] Mechanical, St. Joseph's College of Engineering, Chennai, Tamilnadu, India

*E-mail: karthikeyankmb@stjosephs.ac.in

Abstract. The widespread use of aluminium alloys in the automotive, aerospace, and structural sectors necessitates the establishment of exploratory selection criteria to harness optimality at the various stages of product development. This study utilizes multi-criteria decision-making approaches to determine the most suitable aluminium alloy for thermal friction drilling. Additionally, the process optimization was done using the Taguchi super ranking concept (T-SRC), and the confirmation test using explicit dynamics simulation using ANSYS software. The ideal material selection was evaluated based on several factors, including yield strength, tensile strength, thermal conductivity, impact resistance, density, specific heat capacity, coefficient of thermal expansion, and percentage elongation. The best-ranked AA2024-T4 aluminium alloy underwent an integrated Taguchi Design of Experiments and optimization using Pareto ANOVA and subsequently ANSYS Explicit Dynamics process simulation. The MCDM Expert System derived from AHP, and ranking of materials using TOPSIS, EDAS, and VIKOR proved to be an effective material selection strategy from a set of alternatives for the thermal friction drilling process. By combining Multi-Criteria decision making (MCDM) with process simulation, material selection concerns in different industrial processes can be resolved, therefore eliminating the need for laborious tests.

1. Introduction

Aerospace, automotive, and transportation industries extensively utilize aluminium alloy components. Choosing the right material for various metal manufacturing methods yields components that have excellent strength, low weight, and are cost-effective. It is necessary to choose a suitable material, which was traditionally human-dependent. MCDM techniques are commonly used to find the best solutions for complex problems that involve conflicting objectives, various criteria, interdependent classification responses, and uncertainty in determining precise functional necessities [1]. The resolution of complicated challenges in materials selection necessitates the utilization of conflict management techniques and the consideration of many objectives. These approaches aim to achieve near-optimal solutions by effectively translating functional and cost-benefit criteria. The material selection process holds significant importance in developing products and processes. The improper selection might lead to failure, which can have severe consequences for the end user. These techniques have replaced the reliance on human experts to make decisions that compromise social, economic, and environmental considerations. The objective of this investigation is to identify a method for choosing the most suitable material for a manufacturing process from a range of options. A statistically relevant expert system was formulated using the AHP method. The material properties that affect the manufacturing process are considered to derive the expert system. Decision-making challenges involve the consideration of several goals and standards to make selections.

Nomenclature	
MCDM	Multi-Criteria Decision Making
TFD	Thermal Friction Drilling
AHP	Analytic Hierarchy Process
TOPSIS	Technique for Order Preference by Similarity to Ideal Solution
EDAS	Evaluation Based on Distance from Average Solution
VIKOR	Vlse Kriterijumska Optimizacija I Kompromisno Resenje in Serbian
SRC	Super Ranking Concept
CTE	Co-efficient of Thermal Expansion
ANOVA	Analysis of Variance
ID	Identification
σ_y	Yield Strength, MPa
UTS	Ultimate Tensile Strength, MPa
K	Thermal Conductivity, W/m.K
ρ	Density, g/cm^3
c	Specific Heat, J/g°C
E	Modulus of Elasticity, MPa

The criteria for selecting can be classified according to thermo-mechanical properties, as well as application-specific requirements including wear, fatigue, corrosion, and additional characteristics to consider include the expense of materials, the ability to be recycled, environmental impact, degrees of toxicity, ease of machining, availability of resources, safety, visual appeal, energy usage, and carbon emissions [1]. The material is evaluated using a finite element method based on explicit dynamics simulation, specifically the ANSYS tool.

MCDM approaches are guided by decision model rules or expert systems tailored to individual problems. The challenges arise from the intricate nature of the problems, which are linked to various alternatives with distinct parameters, inadequate data, vagueness, and fuzziness characteristic in real decision-making procedures. Additionally, different algorithms for simultaneously analyzing attributes and substitutions, assumptions, criteria ranking, and the complications in selecting alternatives further contribute to the difficulties encountered. The discrepancies in decision-making that arise from using different weights in various MCDM approaches were highlighted by S.H. Zanakis et al. [2] in their research.

These algorithms may utilize several methodologies to obtain the most optimal solution. The decision-making is significantly impacted by the quantity, caliber, and precision of the information gathered [3]. This was done even when there were competing criteria and constraints. Implementing Multiple Criteria material selection problems, considering economic criteria for various applications that involve complicated or multifunctional material needs, was thoroughly examined [4]. The optimal formulation of piston material was determined in a study conducted by S. Dev et al. [5] using an economic entropy-VIKOR approach.

The investigation aimed at enhancing the mechanical strength while minimizing tribological characteristics, without explicitly conducting experimental assessments [5]. The study, [6] utilized the Fuzzy-TOPSIS approach to find the optimal abrasive material for enhancing surface quality. Additionally, they suggested employing this method for situations when the weights assigned to criteria are ambiguous and inaccurate. For MCDM techniques, the material ranking is distinct and diverse, providing clear-cut choices in general materials selection issues. The Taguchi-SRC eliminates the need for applying weight criteria and predicts ideal parameters for numerous outputs. The Taguchi-SRC approach was used to determine the optimum laser melting settings [7], for analyzing the melting of Inconel 625 nickel alloys. Several studies were conducted to assess the effectiveness of MCDM approaches in optimizing machining operations with many objectives [8-16]. Idapalapati et al. (17) identified several desirable features for metal forming applications, including compressive strength, ductility, wear resistance, and surface hardness. Thermal friction drilling, commonly referred to as green drilling, is a sustainable

IOP Conf. Series: Materials Science and Engineering **1316** (2024) 012014 doi:10.1088/1757-899X/1316/1/012014

technique that is regarded as a viable substitute for traditional metal drilling methods. Friction drilling possesses several technological qualities that provide advantages over conventional drilling methods including fittings, as well as other sheet joining techniques. The process involves the formation of a bushing to form threads after drilling. The bushing material hardens, resulting in increased thread strength. The process does not require cleaning and lubricating agents, making it more technologically and environmentally friendly. There are no chips, which saves material. The process has a high rate and efficiency [18]. Green drilling is environmentally favorable compared to conventional drilling methods since it consumes less power [19].

2. Methodology

This study aims to ascertain the most suited aluminium alloy materials for thermal friction drilling operations from a selection of six possible options. The expert system for evaluating suitable aluminum alloy material (refer to Table 1) for the thermal friction drilling process utilizes probable decision criteria including yield strength, ultimate tensile strength, density, thermal conductivity, specific heat, Young's modulus, CTE, and ductility of the material. The AHP was employed to assess the weights of the criteria. The appropriate material suitable for thermal friction drilling was ranked using TOPSIS, EDAS, VIKOR, and TSRC methods. The L9 orthogonal array proposed by Taguchi was employed for conducting simulations of thermal friction drilling using ANSYS Explicit Dynamics thermo-mechanical analysis as depicted in Figure 1.

Table 1. Prospective Aluminium Alloy Material for Green Drilling Operation [1]

ID	Material	σ_y (MPa)	UTS (MPa)	K (W/m.K)	ρ (g/cm^3)	c (J/g°C)	E (MPa)	CTE	% Ductility
M1	AA2024	324	469	121	2.78	0.875	73.1	23.2	19
M2	AA6063	90	172	200	2.7	0.9	68.9	23.4	22
M3	AA6070	170	320	180	2.71	0.891	69	23.2	20
M4	AA6351	152	248	170	2.71	0.89	68.9	23.4	20
M5	AA2011	125	275	152	2.82	0.864	70	23.1	16
M6	AA2017	275	427	134	2.79	0.88	72.4	23.6	22

The investigative variables are tool rotational speed (rpm), and cutting feed, (mm/min). The optimization process was conducted with the Taguchi-SRC method and subsequently validated through ANSYS Explicit Dynamics thermos-mechanical analysis for the aluminium alloy material that obtained the highest ranking. This investigation focuses on determining the maximum bush height (Bh, mm) and tool contact temperature (T°C) for the best-ranked aluminium alloy AA2024 during thermal friction drilling. The sheet thickness considered for the explicit dynamic analysis is 2 mm. The investigative procedure for selecting the appropriate aluminium alloy material suitable for the manufacturing process under investigation, the green drilling method is illustrated in Figure 1. The thermo-mechanical material parameters are considered for initial scrutiny and assigning weights statistically considering the relevance of the property in enhancing the green drilling performance. The optimal material selection involving simulation, Taguchi SRC, and MCDM paves the way for eliminating numerous experiments, and otherwise uneconomical.

IOP Conf. Series: Materials Science and Engineering **1316** (2024) 012014 doi:10.1088/1757-899X/1316/1/012014

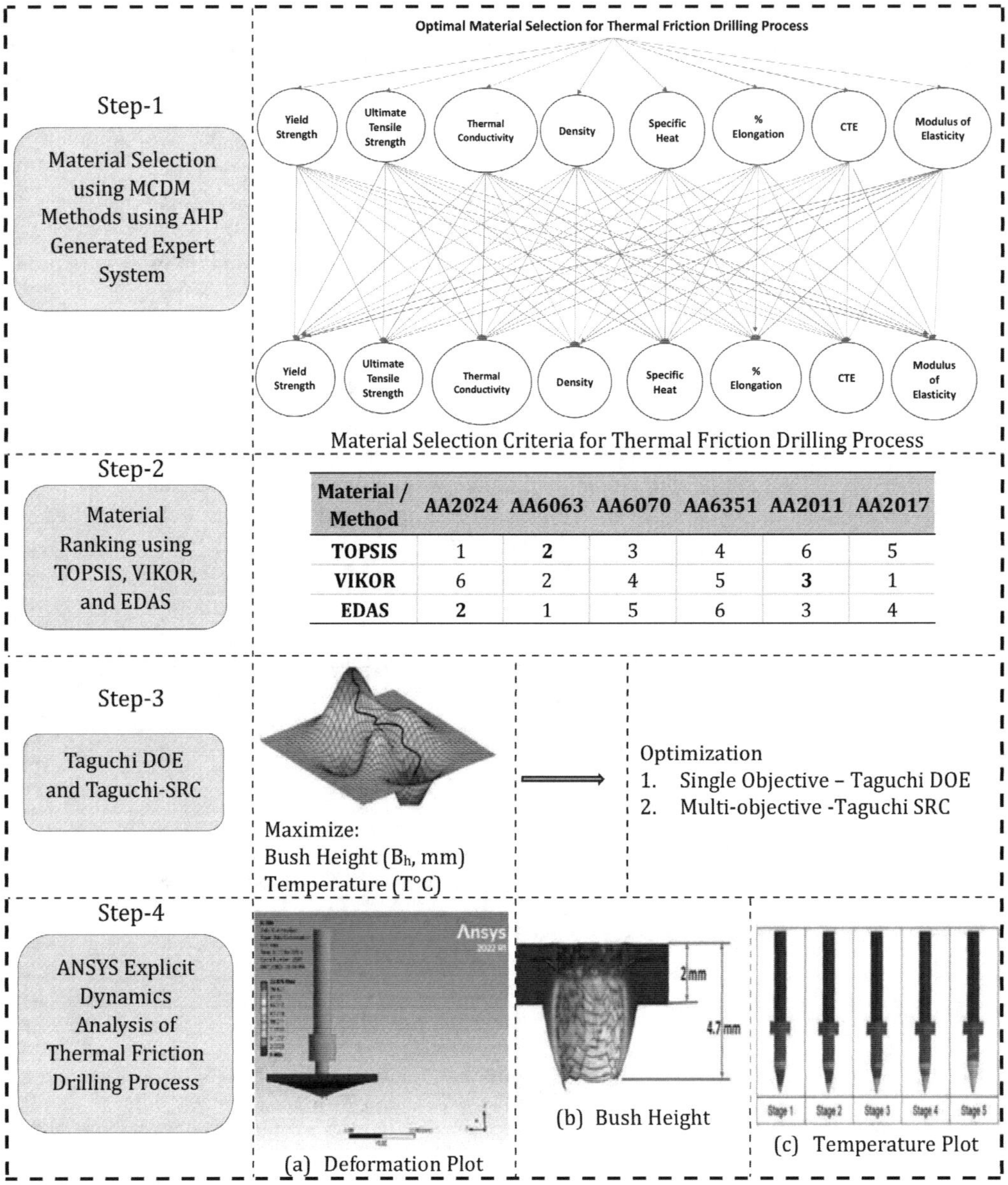

Figure 1. Methodology for Material Selection and Optimization for Green Drilling of Aluminium Alloys

3. Results and Discussions

During the material selection problem decision-making process, there exist M material choices that are evaluated based on N deciding circumstances known as criteria. The options are represented as Mi (where i ranges from 1 to K), while the criteria are designated as Cj (where j ranges from 1 to N). The M1 represents AA2024-T4, whereas M2, M3, M4, M5, and M6 represent alternative aluminium alloy materials [1]. The variable C1 represents the yield strength, while the variables C2, C3, C4, C5, C6, C7, and C8 depict the UTS, K, ρ, c, E, CTE, and % elongation, respectively [1], under T4 condition (refer Table 1). The evaluation criteria for various MCDM methods are primarily determined by the relative significance of substitutes and their impact on attaining a process response close to optimal. In this scenario, the most suitable material for thermal friction drilling is based on the bush height (B_h, mm) and tool contact temperature (T°C).

3.1 Analytical Hierarchy Process (AHP)

Table 2 presents the weighted criteria and rating system for selecting potential aluminium alloy for the thermal friction drilling operation. The rating system is based on a pairwise comparison matrix.

Table 2. The Criteria Weights Expert System using AHP for AAXXX-T4 Aluminium Alloys

Criteria	σ_y (MPa)	UTS (MPa)	K (W/m.K)	ρ (g/cm³)	c (J/g°C)	E (MPa)	CTE	% Ductility	Weights (Wij)
σ_y (MPa)	1	5	6	2	6	4	8	0.5	0.236
UTS (MPa)	0.2	1	0.33	0.2	3	0.5	3	0.2	0.055
K (W/m.K)	0.17	3	1	0.25	3	0.33	5	0.17	0.073
ρ (g/cm³)	0.5	5	4	1	4	2	7	0.2	0.149
c (J/g°C)	0.17	0.33	0.33	0.25	1	0.25	0.5	0.14	0.029
E, (MPa)	0.25	2	3	0.5	4	1	6	0.25	0.101
CTE	0.13	0.33	0.2	0.14	2	0.17	1	0.13	0.030
% Elongation	2	5	6	5	7	4	8	1	0.334

The intricate MCDM problem is transformed into an ordered system to facilitate decision-making. This network incorporates relative criteria and sub-criteria to assess the various possibilities inside the hierarchy network. The initial stage of the AHP is determining the decision hierarchy. Subsequently, the weights for each criterion are determined, and a pairwise judgment matrix is generated. The third phase entails the amalgamation of conclusions, while the ultimate stage involves assessment and verification of the coherence of those judgments [24-25]. A matrix denoted as K x N is created, where K represents several possibilities and N represents multiple criteria. Table 3 presents the evaluation of alternative materials according to the consistency index (CI) and consistency ratio (CR).

Saaty et al. [20-23] advocate using the pair-wise comparison matrix to compare the relative importance of each criterion to other decision-making factors. Subsequently, it is necessary to produce a relatively standardized pairwise matrix. Furthermore, the criteria-weighted matrix is constructed using the assigned scores for various deciding factors. After developing consistency vectors, the following step is to find their major eigenvalue, λ, and the consistency vector. The AHP approach yielded a consistency ratio (CR) of 0.08 ≥ 0.10 for the created decision matrix. The analysis of the AHP computation reveals that the given criteria weightage for choosing the ideal aluminium alloy effectively addresses the intended objective. From a metallurgical standpoint, the formability of sheet metal during the thermal friction drilling operation is primarily influenced by the % elongation, which enhances the material's ability to bear loads during deformation. The decision matrix produced by AHP prioritizes yield strength and density after % elongation as the major criterion.

Table 3. Consistency Ratio (CR) of the Expert System

Criteria	Weights Wij	Consistency Value	Ranking
σ_y (MPa)	**0.236**	**9.431**	**2**
UTS (MPa)	0.055	8.178	6
K (W/m.K)	0.073	8.823	5
ρ (g/cm^3)	0.149	9.587	3
c (J/g°C)	0.029	8.202	8
E (MPa)	0.101	9.290	4
CTE	0.030	7.657	7
% Elongation	**0.334**	**9.307**	**1**

Friction drilling is a process that employs a revolving conical hard tool, which is non-consumable, to establish contact with the workpiece [18-19]. When the contact occurs, the material is heated to a plastic condition as a result of the frictional force. Moreover, because of the axial force, the tool is capable of penetrating the workpiece, leading to the formation of a hole and a bushing. The bush height generated is 2–4 times the sheet thickness. Following this, the tool emerges from the tube in the opposite orientation as depicted in Figure 2. The bushing exhibits a greater thickness compared to the workpiece, leading to an increased usable depth of the thread.

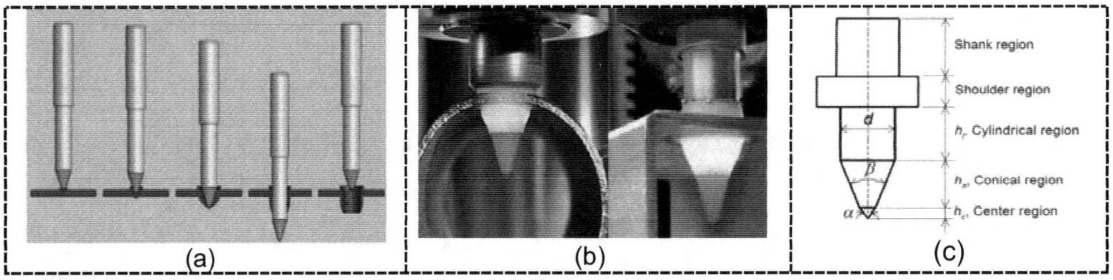

Figure 2. Schematics of Thermal Friction Drilling Process (a) Stages of Thermal Friction Drilling Operation [27] (b) Thermal drilling of tube and sheet metal [28] (c) Friction Drill Tool Geometry [29]

3.2 Technique for Order of Preference by Similarity to Ideal Solution (TOPSIS)

The method finds the best substitute far from the ideal negative solution and near the ideal positive response [30]. The ideal positive solution refers to the set of criteria values that are very desirable, whereas, the optimal negative solution refers to the set of criteria values that are not easy to accomplish [30-33]. The process is to prioritize alternatives that closely resemble the ideal positive solution. It unequivocally differentiates the selected best answer from the substitutes. A selection matrix, Xij, is referred to by the user; and is used to compare the efficacy of multiple options according to predetermined criteria. The assessment is done by assigning weights, denoted as Wij, to each criterion (refer to Table 4). The next step is to determine the positive ideal solution (A+) and negative ideal solution (A-) by considering the categorization of criteria into advantages and costs [34]. The alternative Ai's proximity to A+ is computed. Table 5 displays the best options, where a high proximity value to the perfect answer indicates a good fit. Refer to Table 2 for the rating system for the different properties of the selected aluminium alloy materials.

IOP Conf. Series: Materials Science and Engineering **1316** (2024) 012014 doi:10.1088/1757-899X/1316/1/012014

Table 4. Normalized Weight-Matrix with Optimal Positive and Negative

ID	σ_y (MPa)	UTS (MPa)	K (W/m. K)	ρ (g/cm³)	c (J/g°C)	E (MPa)	CTE	% Ductility
M1	0.15	0.03	0.02	0.06	0.01	0.04	0.01	0.13
M2	0.04	0.01	0.04	0.06	0.01	0.04	0.01	0.15
M3	0.08	0.02	0.03	0.06	0.01	0.04	0.01	0.14
M4	0.07	0.02	0.03	0.06	0.01	0.04	0.01	0.14
M5	0.06	0.02	0.03	0.06	0.01	0.04	0.01	0.11
M6	0.13	0.03	0.02	0.06	0.01	0.04	0.01	0.15
Ideal Positive and Ideal Negative Solution								
A+	0.04	0.03	0.02	0.06	0.01	0.04	0.01	0.15
A-	0.15	0.01	0.04	0.06	0.01	0.04	0.01	0.11
Criteria Weights,	0.236	0.055	0.073	0.149	0.029	0.101	0.030	0.334

Table 5. TOPSIS-based Preference List

Material	Separation Measure		Relative Closeness Value	Rank
	Si+	Si-	Pi	
M1	0.032	0.184	0.851	2
M2	**0.025**	**0.203**	**0.891**	**1***
M3	0.043	0.182	0.810	4
M4	0.037	0.186	0.835	3
M5	0.047	0.163	0.778	5
M6	0.087	0.171	0.663	6

* TOPSIS method prefers AA6063-T4 material for the Thermal Friction Drilling Operation

3.3 Vlsekriterijumska optimizacija Ikompromisno Resenje (VIKOR)Method

The VIKOR index (Q_i) is calculated by assigning weights to utility and regret measurements for different criteria, as presented in Table 6. The AA2014-T4 material outranks others.

Table 6. Utility Measure (S_i) Regret Measure (R_i) VIKOR index (Q_i)

ID	σ_y (MPa)	UTS (MPa)	K (W/m. K)	ρ (g/cm³)	c (J/g°C)	E (MPa)	CTE	% Ductility	Si	Ri	Qi
M1	0.24	0.00	0.000	0.099	0.009	0.000	0.006	0.167	0.52	0.24	0.46
M2	0.00	0.06	0.073	0.000	0.029	0.101	0.018	0.000	0.28	0.10	0.00
M3	0.08	0.03	0.054	0.012	0.022	0.099	0.006	0.111	0.41	0.11	0.19
M4	0.06	0.041	0.045	0.012	0.021	0.101	0.018	0.111	0.41	0.11	0.19
M5	**0.03**	**0.03**	**0.028**	**0.149**	**0.000**	**0.075**	**0.001**	**0.334**	**0.66**	**0.33**	**0.75**
M6	0.19	0.00	0.012	0.112	0.013	0.017	0.030	0.000	0.38	0.19	0.23

**According to VIKOR, AA2011-T4 material is suitable for thermal friction drilling operation*

The robust VIKOR technique shows a strong correlation between characteristics and decision-making factors. After establishing the selection matrix, the next step involves calculating the standardized decision matrix (f_{ij}). After determining the expected results, and computing the utility (Si) or regret (Ri) measure, the decision criterion is categorized as useful [12].

3.4 Evaluation Based on Distance from Average Solution (EDAS)

To optimize uncertainty in real-time stochastic issues, EDAS is used. The attractiveness of alternatives [35] is determined by their distances from the average solution (AVj) and their normal distribution-based performance (Xij). The deviation from the average, both positive and negative, was calculated. The evaluation score (Asi) will be computed based on the normalized weighted sum (NSPi and NSNi). The optimal resolutions are achieved through the utilization of this approach. The outcomes are displayed in Table 7.

Table 7. Weighted Sum (SP/SN), Normalized Weighted Sum (NSP/NSN), and Appraisal Scores (AS)

Material	SPi	SNi	NSPi	NSNi	ASi	Rank
M1	**0.048**	**0.184**	**0.292**	**1.000**	**0.646**	**1***
M2	0.164	0.046	1.000	0.253	0.627	2
M3	0.030	0.012	0.180	0.063	0.122	6
M4	0.052	0.019	0.315	0.106	0.210	5
M5	0.085	0.076	0.517	0.415	0.466	4
M6	0.070	0.109	0.426	0.596	0.511	3

**According to EDAS, AA2024-T4 material is suitable for thermal friction drilling operation*

3.5 Optimal Material Selection Using Preferential Ranking Method

Table.8 shows the ranks of alternate materials using different MCDM approaches. Table.9 shows alternate material rank frequency and membership degrees. Final rank indices and alternate material ranks are in Table.10. The rank index indicates that AA2024-T4 aluminium alloy is the most favorable material for thermal friction drilling operations.

Table 8. Ranking Alternate Materials (AAXXXX-T4) Using Various MCDM Methods

Method / Material	**M1**	**M2**	M3	M4	**M5**	M6
TOPSIS	2	1	4	3	5	6
VIKOR	2	6	4	5	1	3
EDAS	1	2	6	5	4	3

Table 9. Rank Frequency of Alternate Materials (AAXXXX-T4)

Material / Rank	Rank 1	Rank 2	Rank 3	Rank 4	Rank 5	Rank 6
M1	**1**	2	0	0	0	0
M2	**1**	1	0	0	0	1
M3	0	0	0	2	0	1
M4	0	0	1	0	2	0
M5	**1**	0	0	1	1	0
M6	0	0	2	0	0	1

Table 10. Preferential Ranking of Alternate Materials (AAXXXX-T4)

Material / Rank	Rank 1	Rank 2	Rank 3	Rank 4	Rank 5	Rank 6	Final Rank Indices	Final Rank
M1	**0.333**	**0.667**	**0**	**0**	**0**	**0**	**1.667**	**1***
M2	0.333	0.333	0	0	0	0.333	2.997	2
M3	0	0	0	0.667	0	0.333	4.666	6
M4	0	0	0.333	0	0.667	0	4.334	5
M5	0.333	0	0	0.333	0.333	0	3.33	3
M6	0	0	0.667	0	0	0.333	3.999	4

3.6 Single-objective Optimization-Taguchi DOE

The Taguchi L9 DOE investigated AA2024-T4 aluminium alloy simulation output data, including bush height (B_h, mm) and tool contact temperature (T°C). Pareto ANOVA was used to determine TFD variables' contributions and predict output levels as depicted in Table 12. The Super Ranking Concept (SRC) optimizes several outputs and calculates a single factor composition and percentage contribution. The Taguchi L9 matrix design was utilized to gather input-output data for the thermal friction drilling process simulation using ANSYS Explicit Dynamics Analysis. A study was conducted using the "higher the better" equation to determine ideal levels for achieving improved bush height (B_h) and tool contact temperature (T°C) using the Pareto ANOVA statistical approach.

The Pareto ANOVA is conducted to determine the proportionate impact of each factor on the response variables of bush height and tool contact temperature. The findings are displayed in Table 11. The Pareto ANOVA analysis has determined that the A3 (3500 rpm) B3 (125 mm/rev) combination of parameters yields the most favorable results for achieving a greater bush height (B_h, mm), and higher tool contact temperature (T°C). The rotational speed has a substantial impact on the creation of the bush, as seen in Table 12. The height of the bush formation increases proportionally with the increase in rotational speed. Increasing tool velocity during aluminium alloy drilling raises tool-workpiece contact temperature.

Table 11. Pareto ANOVA for Bush Height and Tool Contact Temperature

Factors	Pareto ANOVA for Bush Height				Pareto ANOVA for Tool Contact Temperature			
	Levels	Speed, rpm	Feed, mm/rev	Total	Levels	Speed, rpm	Feed, mm/rev	Total
Sum of Factor Levels	1	40.13	43.03	132.18	1	138.81	143.47	333.73
	2	43.41	44.02		2	43.42	144.54	
	3	**48.63**	**45.13**		3	**151.51**	**147.06**	
Sum of Square of Differences		155.46	9.82	165.29		9421.83	26.91	9448.74
Degrees of Freedom		2	2	4		2	2	4
Percentage Contribution Ratio		0.941	0.059	1		0.995	0.005	1
Cumulative Contribution Ratio		0.941	1			0.995	1	
Optimum Levels		**A3**	**B3**			**A3**	**B3**	

Table 12. ANOVA Results for Bush Height and Tool Contact Temperature

Source	ANOVA for Bush Height, Bh, mm					ANOVA for Tool Contact Temperature, T °C				
	DF	Adj SS	Adj MS	F-Value	P-Value	DF	Adj SS	Adj MS	F-Value	P-Value
Speed, rpm	2	5.002	2.501	140.69	0.000	2	2548.7	12742	165.8	0.000
Feed, mm/rev	2	0.282	0.141	7.94	0.041	2	2264	1132	14.7	0.140
Error	4	0.071	0.017			4	307.3	76.8		
Total	8	5.355				8	28056			
Model Summary						**Model Summary**				
	S	R-sq	R- sq (adj)	R- sq (pred)		S	R-sq	R- sq (adj)	R- sq (pred)	
	0.133	98.67%	97.34%	93.28%		8.76	98.9%	97.81%	94.45%	

3.7 Multi-objective Optimization-Taguchi DOE and Taguchi-SRC

The materials chosen using MCDM approaches are evaluated through finite element simulation using ANSYS Explicit Dynamics thermos-mechanical analysis. Tables 13 and 14 exhibit the Taguchi DOE and Taguchi-SRC variables for conducting the simulation study using the ANSYS software, respectively. The height of the bush (B_h, mm) and the temperature at the point of contact with the tool (T°C) are the outcomes of the simulation study. In this case, the criteria for the optimum parameter combination and the suitable material for the thermal friction drilling process are based on the "larger the better hypothesis". The simulation results for the bush height (B_h, mm) and the tool contact temperature (T°C) are shown in Table 14. The parameters that have been determined to be the most effective are the rotational speed, 3500 rpm, and the feed rate, 125 mm/rev.

Table 13. Taguchi DOE Plan

Control Factors	Level-1	Level-2	Level-3
Speed, N, rpm	2500	3000	3500
Feed, mm/rev	75	100	125

Table 14. Thermal Friction Drilling Parameters and L9 Orthogonal Array Levels

Test Run	Speed N, rpm	Feed, mm/rev	Bush Height B_h (mm)	Tool Temperature , T°C	SN Ratio		Rank		Squared Rank		Sum of Squared Rank, SSR	Super Ranking, SR
					B_h	T°C	B_h	T°C	B_h	T°C		
1	2500	75	4.5	199	13.06	45.98	9	9	81	81	162	9
2	2500	100	4.7	200	13.44	46.02	8	8	64	64	128	8
3	2500	125	4.8	219	13.62	46.81	7	7	49	49	98	7
4	3000	75	5	241	13.98	47.64	6	6	36	36	72	6
5	3000	100	5.2	251	14.32	47.99	5	5	25	25	50	5
6	3000	125	5.7	286	15.12	49.13	4	4	16	16	32	4
7	3500	75	6.3	311	15.99	49.86	3	3	9	9	18	3
8	3500	100	6.5	336	16.26	50.53	2	2	4	4	8	2
9	**3500**	**125**	**6.6**	**360**	**16.39**	**51.13**	**1**	**1**	**1**	**1**	**2**	**1***

*Optimal process parameter: Speed, 3500 rpm; Feed, 125 mm/rev for thermal friction drilling of AA2024-T4 Material

3.8 ANSYS Explicit Dynamics -Thermal Friction Drilling Simulation

The simulation results of the green drilling operation are displayed in Figure 2. The height of the bush (B_h, mm) and the temperature at the point of contact with the tool (T°C) is determined by simulating various test settings as mentioned in Table 14 using ANSYS software as the response variable. The identified optimal process parameters for thermal friction drilling of AA2024 aluminium alloy from process simulation are as follows: speed of 3500 rpm, feed of 125 mm/rev. The maximum height of the bush, without including the petals shown in Figure 3(d-f), is 6mm. This height is suitable for the subsequent tapping operation.

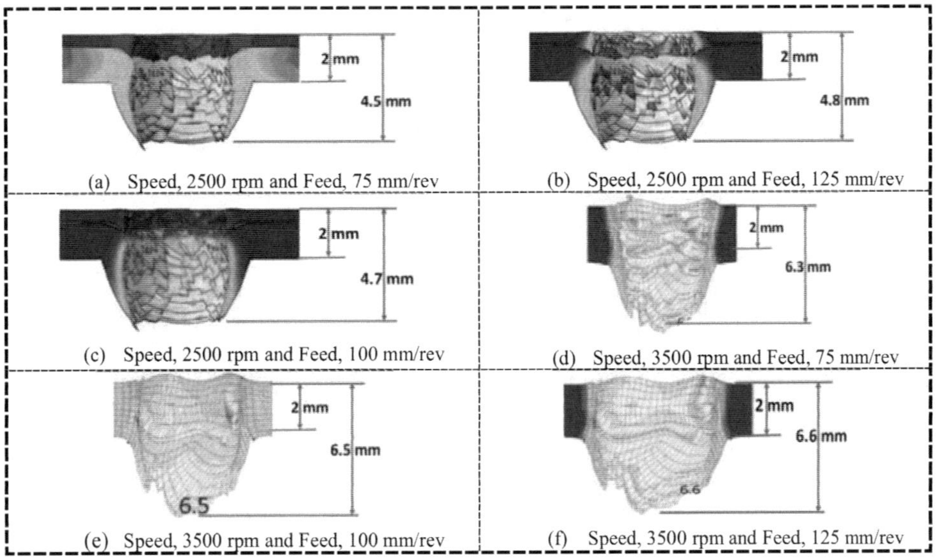

Figure 3. Representative ANSYS Explicit Dynamics Simulation Deformation Plots for AA2024 Aluminium Alloy depicting Bush Height, B_h, mm

Conclusion

This research set out to identify potential aluminium alloys for thermal friction drilling. The AHP approach was used to determine the decision matrix and weighting for the various material attributes. The MCDM methods, such as TOPSIS, VIKOR, and EDAS, were employed to determine the optimal alternate material by computational analysis. Subsequently, a process-oriented simulation was performed using ANSYS Explicit Dynamics to analyze the thermal friction drilling operation of the highest-rated AA2024-T4 aluminium alloy. The conclusive outcomes are listed below;

- AA2024-T4 aluminium alloy was chosen as the prospective material for thermal friction drilling operations. The EDAS approach demonstrated a high level of accuracy in forecasting outcomes, comparable to that of the TOPSIS and VIKOR methods.
- The AHP approach accurately determines the contribution of material characteristics for achieving the combination of material attributes that improve the thermal friction drilling process. The percentage elongation and yield strength of the material are important factors that significantly impact the effectiveness of thermal friction drilling.
- The rank frequency index indicates that AA2024-T4 aluminium alloy is the most suitable material for thermal friction drilling operations compared to other aluminium alloys.
- A simulation was conducted using ANSYS Explicit Dynamics. The effective bush height (Bh) from the simulation is 2.5X the sheet thickness. The response from the simulation is used to run the Taguchi-SRC to find the ideal process parameters for the superior alternative material, AA2024-T4 aluminium alloy.
- The following parameters have shown to be the most effective: There is a feed rate of 125 mm/rev and a rotational speed of 3500 rpm.

- The uncertainty in the MCDM process for picking the optimal material in green drilling operations using different approaches is apparent. Therefore, it is necessary to conduct a comparative assessment to determine the criteria for material selection.

The integration of MCDM (Multi-Criteria Decision Making) and Manufacturing Process Simulation enables the resolution of material selection issues in various industrial processes, hence removing the need for time-consuming experiments.

References

[1] A.H. Abishini, K.M.B. Karthikeyan, Application of MCDM and Taguchi super ranking concept for materials selection problem, Materials Today: Proceedings, Volume 72, Part 4, 2023, Pages 2480-2487, ISSN 2214-7853, https://doi.org/10.1016/j.matpr.2022.09.526.

[2] Xenakis, S. H., Solomon, A., Wishart, N., & Dublish, S. Multi-attribute decision making: A simulation comparison of select methods. European Journal of Operational Research 1998; 107(3):507–529. https://doi.org/10.1016/S0377-2217(97)00147-1.

[3] Mardani, A., Jusoh, A., MD Nor, K., Khalifah, Z., Zakwan, N., & Valipour, A. Multiple criteria decision-making techniques and their applications – a review of the literature from 2000 to 2014. Economic Research-Ekonomska Istraživanja 2015; 28(1):516–571. https://doi.org/10.1080/1331677X.2015.1075139.

[4] Emovon, I., & Oghenenyerovwho, O. S. Application of MCDM method in material selection for optimal design: A review. Results in Materials 2020; 7: 100-115. https://doi.org/10.1016/j.rinma.2020.100115.

[5] Dev, S., Aherwar, A. & Patnaik, A. Material Selection for Automotive Piston Component Using Entropy-VIKOR Method. Silicon 2020; 12: 155–169. https://doi.org/10.1007/s12633-019-00110-y.

[6] Maity, S. R., & Chakraborty, S. Grinding Wheel Abrasive Material Selection Using Fuzzy TOPSIS Method. Materials and Manufacturing Processes 2013; 28(4): 408–417. https://doi.org/10.1080/10426914.2012.700159.

[7] Divya, C., Raju, L.S., Singaravel, B. Application of MCDM Methods for Process Parameter Optimization in Turning Process-A Review. In: Narasimham, G.S.V.L., Babu, A.V., Reddy, S.S., Dhanasekaran, R. (eds) Recent Trends in Mechanical Engineering. Lecture Notes in Mechanical Engineering. Springer, Singapore. 2021; 199-207. https://doi.org/10.1007/978-981-15-7557-0_18.

[8] B. Singaravel, T. Selvaraj, Multi criteria optimization using analytic hierarchy process in turning operation. International Journal of Machining and Machinability of Materials 2017; 19(3):218–229. https://doi/abs/10.1504/IJMMM.2017.084005.

[9] A. Gok, A new approach to minimization of the surface roughness and cutting force via fuzzy TOPSIS, multigrey design, and RSA. Measurement 2015; 70: 100–109. https://doi.org/10.1016/j.measurement.2015.03.037

[10] H. Majumder, A. Saha, Application of MCDM based hybrid optimization tool during turning of ASTM A588. Decis. Sci. Lett. 2018; 7: 143–156. https://dx.doi.org/10.5267/j.dsl.2017.6.003.

[11] N.S. Narayanan, N. Baskar, M. Ganesan, M.P. Jenarthanan, S. Praveen, Evaluation and optimization of surface roughness and metal removal rate through RSM, GRA, and TOPSIS techniques in turning PTFE polymers, in Advances in Manufacturing Technology, Springer, Singapore 2019; 595–605. https://doi.org/10.1007/978-981-13-6374-0_65.

[12] D. Palanisamy, P. Senthil, Optimization on turning parameters of 15-5PH stainless steel using Taguchi based grey approach and TOPSIS. Arch. Mech. Eng. 2016; 63: 397–412. https://doi.org/10.1515/meceng-2016-0023.

[13] L.I. Tong, C.C. Chen, C.H. Wang, Optimization of multi-response processes using the VIKOR method. Int. J. Adv. Manuf. Tech. 2007; 31: 1049–1057. https://doi.org/10.1007/S00170-005-0284-6.

[14] R. Thirumalai, J.S. Senthilkumar, Multi-criteria decision making in the selection of machining parameters for Inconel 718. J. Mech. Sci. Tech. 2013; 27:1109–1116. http://dx.doi.org/10.1007/s12206-013-0215-7.

[15] Chakraborty, S., Chakraborty, S. A Scoping Review on the Applications of MCDM Techniques for Parametric Optimization of Machining Processes. Arch Computat Methods Eng. 2022. https://doi.org/10.1007/s11831-022-09731-w.

[16] Majid Behzadian, S. Khanmohammadi Otaghsara, Morteza Yazdani, Joshua Ignatius, A state-of-the-art survey of TOPSIS applications, Expert Systems with Applications 2012; vol.39: 17: 13051-13069, ISSN 0957-4174, https://doi.org/10.1016/j.eswa.2012.05.056.

[17] Idapalapati, S. et al. (2013). Materials in Metal Forming. In: Nee, A. (eds) Handbook of Manufacturing Engineering and Technology. Springer, London. https://doi.org/10.1007/978-1-4471-4976-7_42-1.

[18] Eliseev, A., Kolubaev, E. Friction drilling: a review. Int J Adv Manuf Technol 116, 1391–1409 (2021). https://doi.org/10.1007/s00170-021-07544-y

[19] Pereira O, Urbikain G, Rodrigues A, Calleja A, Ayesta I, Lopez de Lacalle LN (2019) Process performance and life cycle assessment of friction drilling on dual-phase steel. J Cleaner Prod 213:1147– 1156
[20] Saaty, T. L. Theory and Applications of the Analytic Network Process. Pittsburgh, PA: RWS Publications, 4922 Ellsworth Avenue, Pittsburgh, PA 15213-2005.
[21] Saaty T.L. How to make a decision: The Analytic Hierarchy Process. European Journal of Operational Research 1990. https://doi.org/10.1016/0377-2217(90)90057-I.
[22] Saaty, T.L. Decision making for Leaders, RWS Publications, 4922 Ellsworth Avenue, Pittsburgh 1996.
[23] Saaty, T.L., Vargas, L.G. How to Make a Decision. In: Models, Methods, Concepts & Applications of the Analytic Hierarchy Process. International Series in Operations Research & Management Science 2012; 175. https://doi.org/10.1007/978-1-4614-3597-6_1.
[24] E. Triantaphyllou, B. Shu, S. Nieto Sanchez, T. Ray, Encyclopaedia of Electrical and Electronics Engineering, (J.G. Webster, Ed.), John Wiley & Sons, New York, NY 1998; 15: 175-186.
[25] Benitez JM, Martin JC, Roman C. Using fuzzy number for measuring quality of service in the hotel industry. Tour Manag 2007; 28:544-555. https://doi.org/10.1016/J.TOURMAN.2006.04.018.
[26] Kaya, Mehmet & Aktaş, Alaattin & Beylergil, Bertan & Akyildiz, Hamza. (2014). An experimental study on friction drilling of ST12 steel. Transactions- Canadian Society for Mechanical Engineering. 38. 319-329. 10.1139/tcsme-2014-0023.
[27] Kaya, Mehmet & Aktaş, Alaattin & Beylergil, Bertan & Akyildiz, Hamza. (2014). An experimental study on friction drilling of ST12 steel. Transactions- Canadian Society for Mechanical Engineering. 38. 319-329. 10.1139/tcsme-2014-0023.
[28] Shane, Thermal Melting Drilling: Principles and Applications, https://www.machinemfg.com/thermal-melting-drilling/, August 25, 2023.
[29] El-Bahloul, Sara & Elshourbagy, Hazem & Al-Makky, Mohammad & El-Midany, Tawfik. (2013). Thermal Friction Drilling: (A Review).
[30] Ertugrul D, Karakasoglu N. Performance evaluation of Turkish cement firms with fuzzy analytic hierarchy process and TOPSIS methods. Exp Syst Appl Inter J 2009; 36:702-715. https://doi.org/10.1016/j.eswa.2007.10.014.
[31] Wang YM, Elhag TMS. Fuzzy TOPSIS method based on alpha level sets with an application to bridge risk assessment. Exp Syst App Inter J 2006; 31:309-319. https://doi.org/10.1016/j.eswa.2005.09.040.
[32] Wang TC, Chang TH. Application of TOPSIS in evaluating initial training aircraft under a fuzzy environment. Exp Syst App Inter 2007; 33:870-880. https://doi.org/10.1016/j.eswa.2006.07.003.
[33] Wang YJ, Lee HS. Generalizing TOPSIS for fuzzy multiple-criteria group decision-making. Comp Math Appl 2007; 53:1762-1772. https://doi.org/10.1016/j.camwa.2006.08.037.
[34] M.R. Sanjay, Mohammad Jawaid, N.V.R. Naidu, B. Yogesha, 11 - TOPSIS method for selection of best composite laminate, Editor(s): Mohammad Jawaid, Mohamed Thariq, Naheed Saba, In Woodhead Publishing Series in Composites Science and Engineering, Modelling of Damage Processes in Biocomposites, Fibre-Reinforced Composites and Hybrid Composites, Woodhead Publishing 2019;199-209, https://doi.org/10.1016/B978-0-08-102289-4.00011-4.
[35] Mehdi Keshavarz Ghorabaee, Maghsoud Amiri, Edmundas Kazimieras Zavadskas, Zenonas Turskis and Jurgita Antucheviciene. Stochastic EDAS method for multi-criteria decision-making with normally distributed data, Journal of Intelligent & Fuzzy Systems 2017; 33: 1627–1638. http://dx.doi.org/10.3233/JIFS-17184.

IOP Conf. Series: Materials Science and Engineering **1316** (2024) 012015 doi:10.1088/1757-899X/1316/1/012015

Hot Ductility Behaviour of AA7075-T651 Aluminium Alloy

KMB. Karthikeyan*, Cyril Moses L, Eshwar NK, Sam Richard N, Sai Ram B

Mechanical, St. Joseph's College of Engineering, Chennai, Tamilnadu, India

*E-mail: *karthikeyankmb@stjosephs.ac.in*

Abstract. This study focused on examining the hot ductility characteristics of aluminium alloy 7075-T651. Hot ductility is the capacity of a substance to undergo deformation at high temperatures without fracturing. Aluminium alloys typically demonstrate favorable hot ductility, rendering them appropriate for a wide range of high-temperature applications. The investigation entails performing hot tensile tests for AA7075-T651 aluminium alloy at different deformation temperatures (350°C-525°C), followed by fractography to ascertain the fracture mode. The ductility of the AA7075-T651 alloy was greatly enhanced during this procedure. The findings indicate that the hot working range of the aluminium alloy 7075-T651 can be expanded to 525°C instead of a liquidus start temperature of 477°C with a maximum of 30% ductility. The fractography results demonstrate that the fracture mode remains ductile. The AA7075-T651 aluminium alloy exhibits excellent thermal stability, allowing it to be hot worked at temperatures as high as 525°C without experiencing any sudden alterations in its characteristics or fracture behaviour.

1. Introduction

Aluminium alloys are a prominent lightweight material increasingly utilized due to their advantages, such as a favorable strength-to-weight ratio and excellent recyclability. Aluminium alloys possess a combination of lightweight characteristics, high strength, resistance to corrosion, and exceptional thermal and electrical conductivity [1-2]. These qualities render them essential in contemporary industry and diverse technical applications. It possesses the second-highest degree of malleability among all metals and the sixth-highest degree of ductility. Understanding the behaviour of aluminium alloys at different temperatures is crucial for optimizing their formability and mechanical properties. Raising the temperature causes the material to become softer, resulting in a decrease in its strength and an increase in its ductility. Nevertheless, elevated temperature can also result in the enlargement of grain size and a reduction in the ability of a material to deform due to the creation of sizable, rough grains with limited capacity for deformation.

As the temperature rises, the ductility of aluminium alloy sheets increases. As an illustration, the 7075-T4 aluminium alloy sheet demonstrates a rise in elongation from 15% when it is rapidly cooled to 26.1% at a temperature of 440 °C [3]. At higher temperatures, the increased thermal energy enables dislocations to migrate with greater ease within the crystal lattice. The enhanced ability to move allows for plastic deformation through the process of dislocation glide and climb, ultimately increasing ductility [4]. As the temperature increases, the required flow stress for inducing plastic deformation

decreases. The reduction in flow stress allows the material to undergo greater bending and stretching without fracturing, hence improving its ductility [5].
The formability of aluminium alloys can be significantly enhanced at higher temperatures, while their spring back can be reduced. Consequently, hot forming has gained popularity and widespread usage, especially for fabricating complicated components [6]. Even though hot forming conditions can increase the formability of aluminum alloys, softening processes like recrystallization and recovery result in a loss of strength. Therefore, further artificial aging is typically required to increase the applied materials' strength [7].

Hot deformation is the act of shaping or molding aluminium alloys at high temperatures, usually above 0.4 times the melting point of the alloy. This temperature range facilitates more malleability and enhances the ease of shaping the material. Aluminium alloys exhibit plastic deformation at high temperatures, allowing them to be molded without fracturing. As the temperature increases, the material exhibits increased ductility, meaning it can be easily formed without fracturing. The reason for this is that elevated temperatures promote plastic deformation by enhancing atomic mobility inside the material [8]. It is crucial to acknowledge that there exists an ideal temperature range for hot deformation. Insufficient temperature might result in reduced ductility of the material, which hampers its malleability. Conversely, when the temperature is very high, there is a risk of excessive grain growth, which can have a detrimental impact on the mechanical properties of the alloy. Therefore, it is essential to identify the optimal temperature range to achieve accurate hot deformation of aluminium alloys.
The standard temperature range for the process of hot deformation in aluminium alloys falls within the range of 300°C to 470°C [9-11]. This range enables the alloy to achieve the highest level of ductility and formability during the process of deformation. It is crucial to acknowledge that the precise temperature range may differ according to the alloy composition and the required deformation properties. The limited malleability of the aluminium alloy at normal temperatures restricts its usage in the automotive sector, particularly for AA7XXX alloys. The 7000 series aluminium alloys can be used in various applications. Manufacturers and end-users strongly appreciate the exceptional combination of lightweight and toughness exhibited by grade AA7075-T651. It is extensively utilized in the automotive, aircraft, and aerospace industries due to its strength and machinability. Aluminium alloy AA7075-651 aluminium alloy is commonly used for highly stressed parts, including gears, fuse parts, structural components, and bows [12].

The solidus temperature of aluminium alloy AA7075 starts at 477°C, whereas its liquidus temperature is 635°C [13]. This alloy undergoes eutectic melting, transitioning from a solid to a liquid state within the temperature range of 477°C to 560°C. The functioning range of this is limited to 470°C. The features of aluminium alloys, such as their lightweight, corrosion resistance, and exceptional strength, are essential in industries including automotive, aerospace, and construction, where these alloys are frequently utilized [14-15]. This experiment aims to determine the temperature range at which the AA7075-T651 heat-treated aluminium alloy may be effectively hot-worked. In addition, the objective is to assess the mechanical characteristics of the material at different deformation temperatures ranging from 350°C to 525°C, as well as its performance beyond the solidus temperature of 477°C.

2. Materials and Methods

2.1 Hot Tensile Test

The hot tensile test is performed to determine the temperature range at which the sample material can be effectively shaped during forming operations. The research methodology is presented in Figure 1. The optical emission spectroscopy results for the heat-treatable AA7075-T651 aluminium alloy are presented in Table 1. The material was normalized at 415°C for 3 hours.

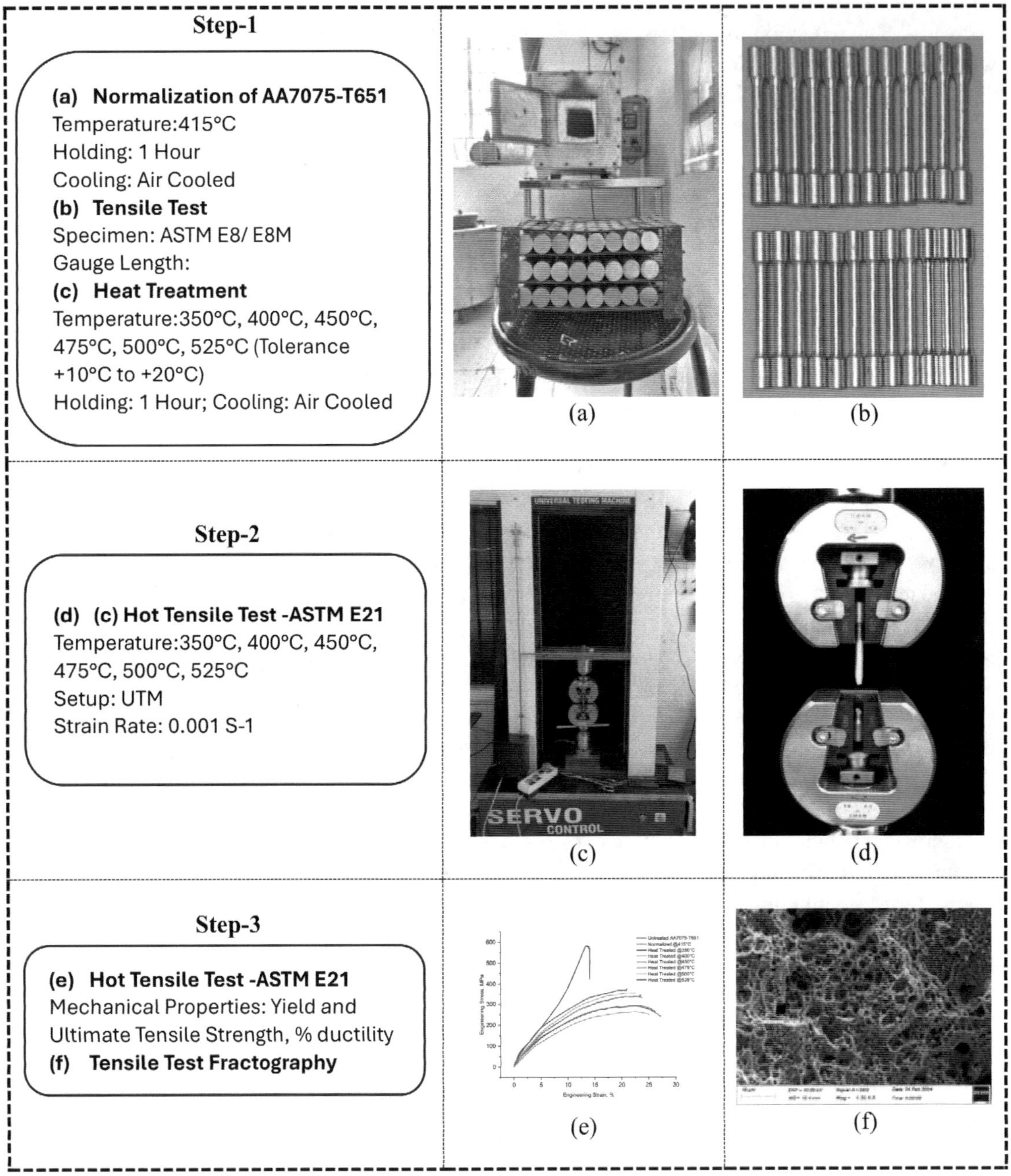

Figure 1. Research Methodology (a) 1000°C Max Electric Box Furnace Setup (b) Tensile Test specimens (c) (d) UniversalTesting Machine for Hot Tensile Test (e) Engineering Stress-Strain Curves for Various Heat Treated AA7075-T651 Aluminium Alloy (f) Tensile Test Fractograph

Then the specimens were fabricated to conform to the standard dimensions for tensile testing specified in the ASTM E8M. The tensile test materials underwent several heat treatments in a standard box furnace, as indicated in Table 2, and were subsequently cooled by air. The heating rate was maintained at a consistent rate of 10°C per minute. Tensile tests were performed using the Fuel Instruments and Engineers, a 5-ton capacity, Universal tensile testing machine following the ASTM-E21 standard [16]. Tensile tests were performed at a tensile rate of 0.001 s–1. The replication of each test was performed three to ensure the precision of the experiment. To determine the failure mode a ZEISS scanning electron microscope was used to perform a tensile test fractography analysis. To determine the hot working range for the examined AA7075-T651 aluminium alloy, the heat treatment conditions were carefully prepared to understand the treatment's effect on the ductility, material strength, and failure modes. Also, to determine the temperature range within which the AA7075-T651 aluminium alloy can be operated.

Table 1. Chemical Composition Results for AA7075-T651 Aluminium Alloy

Elements	Si	Fe	Cu	Mn	Mg	Zn	Cr	Ti	Al
Composition, %	0.19	0.3	1.2	0.14	2.1	5.2	0.17	0.01	Balance

Table 2. Heat Treatment Conditions for AA7075-T651 Aluminium Alloy

Material ID	Heat Treatment Condition
R	Untreated AA7075-T651 Aluminium Alloy
H	Normalized at 415°C for 3 Hours
H1	Heat Treated at 350°C for 1 Hour and Air Cooled
H2	Heat Treated at 400°C for 1 Hour and Air Cooled
H3	Heat Treated at 450°C for 1 Hour and Air Cooled
H4	Heat Treated at 475°C for 1 Hour and Air Cooled
H5	Heat Treated at 500°C for 1 Hour and air-cooled
H6	Heat Treated at 525°C for 1 Hour and Air Cooled

3. Results and Discussion

3.1 Hot Tensile Test

The tensile strength, yield, % ductility, % reduction in necking area, and modulus of toughness are presented in Table 3. From Figure 2(a) and (b), the ultimate tensile strength drops substantially from 580.33 MPa to 278.33 MPa at 350°C. Thermal softening causes the material to become more ductile as the temperature rises from the H2 condition to the H6 condition (range:400°C to 525°C). Following the normalizing procedure, the yield stress decreases from 443.6 MPa to 211.50 MPa at 350°C. A decrease in the yield strength causes an increase in elongation, which subsequently results in a more substantial decrease in the cross-sectional area in the necking region. This verifies that the material is in an elastoplastic condition, and can be used for secondary processing like forming, rolling, and extrusion at a hot working range of 350°C to 525°C. The AA7075-T6 aluminium alloy exhibits excellent thermal stability, allowing it to be effectively hot worked at temperatures up to 525°C without experiencing any sudden alterations in its properties or microstructure (refer to Figure 4). A substantial change in % ductility and % reduction in area was observed from the hot working temperature of 400°C (refer to Figure 3).

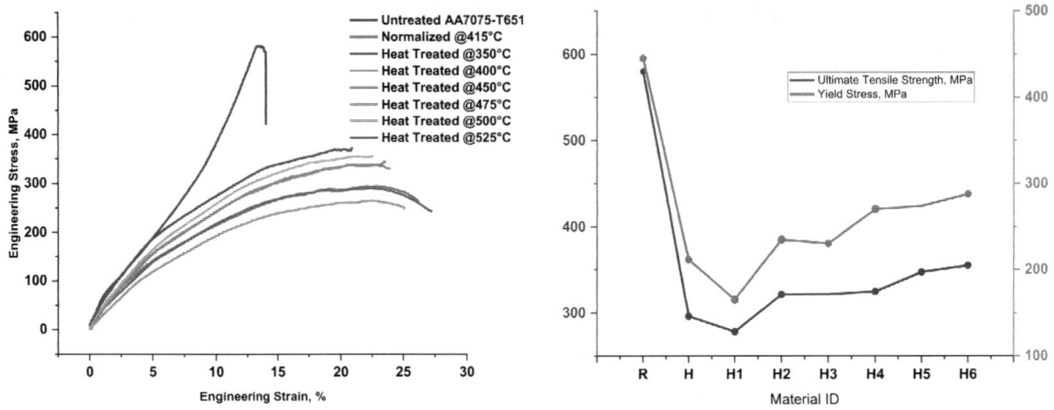

(a) **(b)**

Figure 2. Tensile Test Results for Various Heat Treatment Conditions for AA7075-T651 Aluminium Alloy (a) Engineering Stress Strain Plot (b) Ultimate and Yield Stress Plot

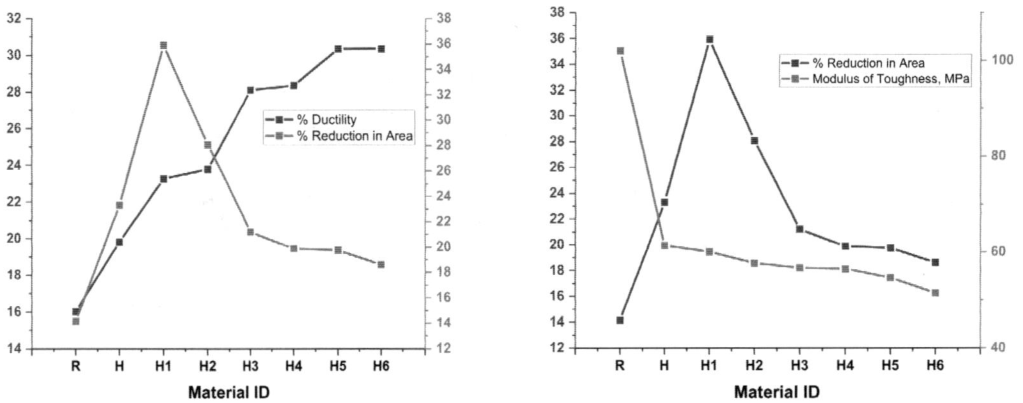

(a) **(b)**

Figure 3. Tensile Test Results for Various Heat Treatment Conditions for AA7075-T651 Aluminium Alloy (a) % Ductility and % Area Reduction Plot (b) Modulus of Toughness Plot

Table 3. Hot Tensile Test Results for Various Heat Treatments for AA7075-T651 Aluminium Alloy

Material ID	Ultimate Tensile Strength, MPa	Yield Strength, MPa	% Ductility	% Area Reduction	Modulus of Toughness, MPa
R	580.33	445.25	16.02	14.15	102.03
H	296.14	211.50	19.82	23.31	61.38
H1	278.33	165.28	23.27	35.91	60.06
H2	321.29	235.20	23.79	28.07	57.70
H3	321.45	230.82	28.11	21.20	56.66
H4	324.66	173.07	28.35	19.87	56.47
H5	347.26	274.16	30.35	19.74	54.65
H6	354.81	287.84	30.36	18.60	51.44

Thereafter, a phenomenal increase in ductility was observed, and the material behavior after the liquidus temperature 477°C up to 525°C was still elastoplastic with a consistent reduction in its toughness (57.70 MPa to 51.44 Mpa) as presented in Figure 3 (b). The Modulus of Toughness experiences a significant decline following the process of normalizing, specifically from the T651 temper condition to the H (350°C) condition. From the temperature range 350°C to 525°C, the decrease is more marginal, and it remains within a range of 50-60 MPa. The drop in toughness of the material is attributed to thermal softening, which causes the material to become more workable. The modulus of toughness of a material is influenced by its ductility[17-18]. If two materials have the same yield strength, their modulus of toughness can differ significantly if one material is more ductile than the other. Ductile materials exhibit more elongation and possess a larger region beneath their stress-strain curve [19].

The temperature at which material deformation occurs can be estimated by considering the desired level of ductility, as revealed by experimental observations. To attain a ductility ranging from 15% to 27%, the appropriate temperature range for hot working of AA7075-T651 should fall within the H and H4 conditions, specifically between 350°C and 475°C. To attain a ductility of 30%, it is necessary to subject the material to a temperature of 525°C while under the H6 condition. Within the temperature range of 400°C to 525°C, the material experiences an increase in ductility as a result of induced elastoplastic activity. This is evidenced by a reduction in the necking region, as depicted in Table 3 and Figure 3(b). The necking region experiences a maximum reduction of 150% under the H1 (350°C) condition, while the H4 (475°C) to H6 (525°C) conditions show a marginal reduction of 40% compared to the untreated AA7075-T651 condition.

3.2 Tensile Test Fractography

The fractography examination of the hot tensile tested samples indicated a ductile fracture with significant dimple formation. The fracture surfaces were examined using scanning electron microscopy and the corresponding images are presented in Figure 4 (a)-(e). Fracture surfaces investigated following the tensile testing showed a prominent presence of dimple development. Additionally, there are visible shear dimples and deep shear dimples on the surface of the crack (refer to Figure 4d-4e). Additionally, the presence of cleavage planes and protrusions [20] was observed for the T651 tempered condition (refer to Figure 4a). The thermal processing of AA7075-T651 under the H4 (475°C) to H6 (525°C) exhibited a fracture characteristic that is either ductile or semi-ductile fracture with cleavage facets surrounded by dimples predominantly [21-23]. The ductility changes considerably and up to 30% for the H6 (525°C) condition.

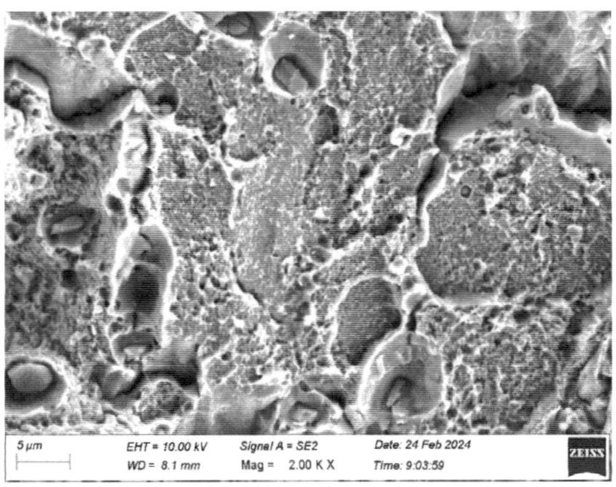

(a) AA7075-T651 Temper Condition

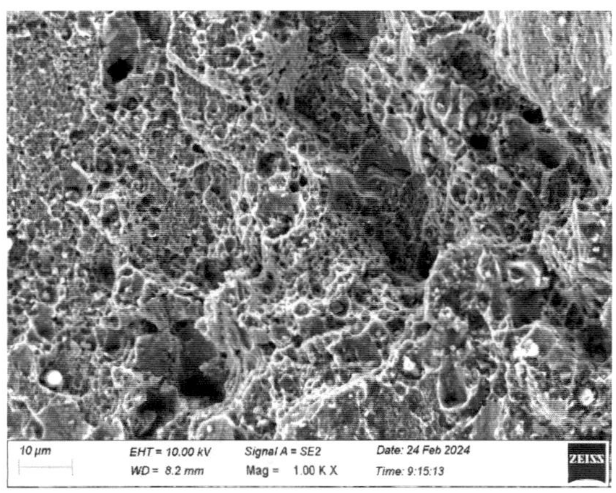

(b) Normalized at 415°C -3 Hours -Air Cooled

IOP Conf. Series: Materials Science and Engineering **1316** (2024) 012015 doi:10.1088/1757-899X/1316/1/012015

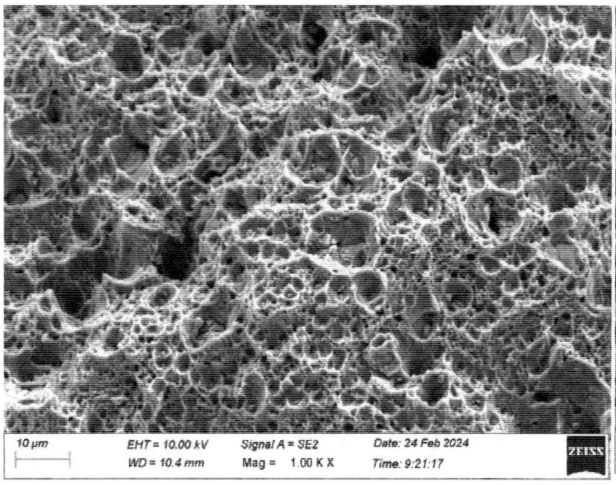

(c) Heat Treated at 350°C-1 Hour-Air Cooled

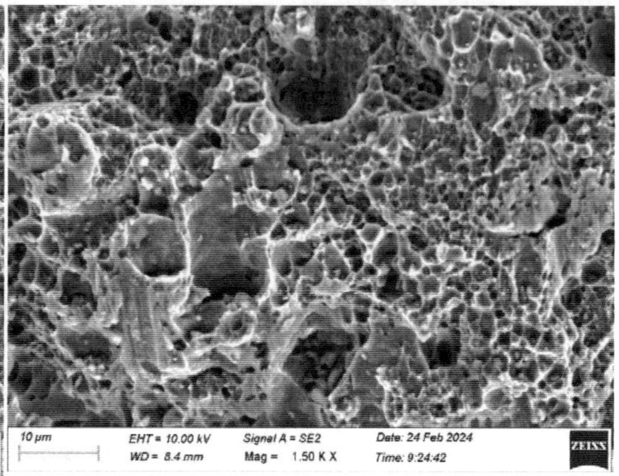

(d) Heat Treated at 475°C-1 Hour -Air Cooled

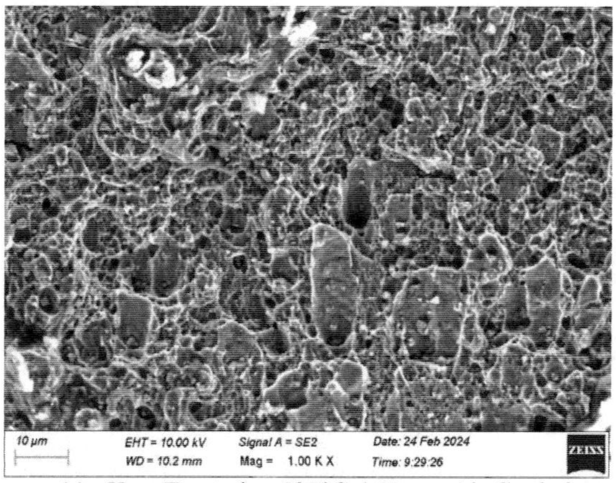

(e) Heat Treated at 525°C-1 Hour -Air Cooled

Figure 4. Tensile Test Fractography SEM Images for Various Heat Treatment Conditions for AA7075-T651 Aluminium Alloy (a) –(e)

3.3 Microstructural Characterization

The microstructure was examined using an optical microscope and the corresponding images are presented in Figure 5 (a)-(c). The specimen is exposed to Weck's reagent for an ideal duration of 8 seconds, guaranteeing thorough saturation and subsequent rinsing with water.

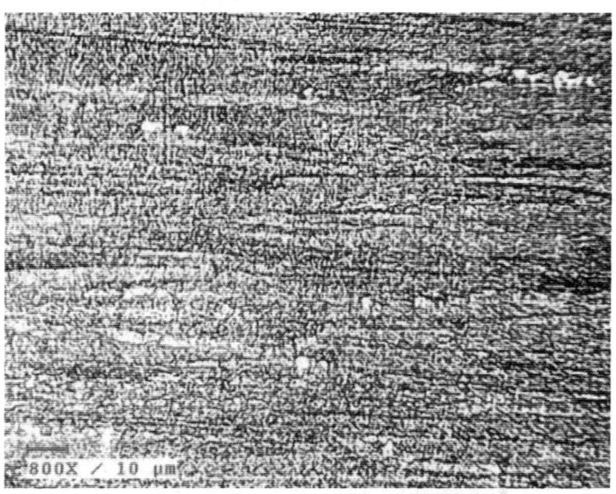

(a) Microsture of AA7075-T651

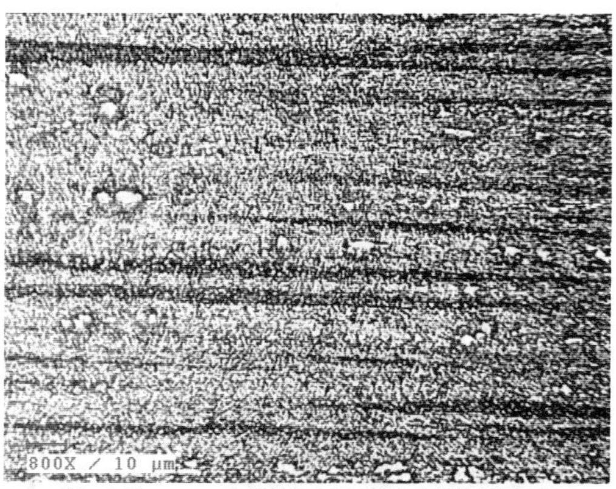

(b) Microsture of Normalized AA7075-T651

IOP Conf. Series: Materials Science and Engineering **1316** (2024) 012015 doi:10.1088/1757-899X/1316/1/012015

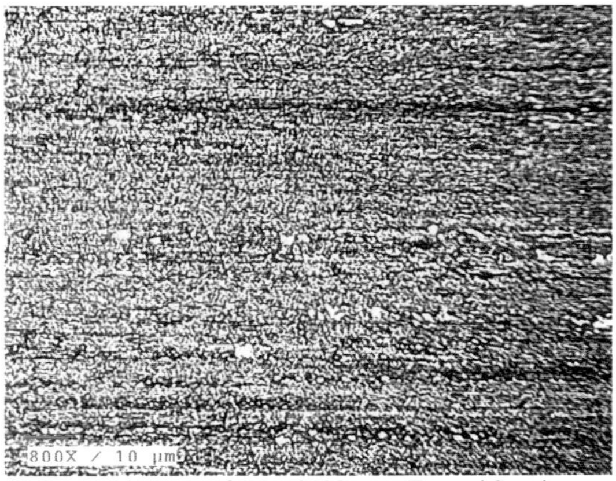

(c) Microsture of H6 -525°C Heat Treated Specimen

Figure 5. Representative Micrographs of AA7075-T651 and Heat-Treated Conditions:(a)–(c)

The specimens were subjected to analysis of their microstructures at various temperatures. The study showed a well-defined equiaxed grain structure, along with the presence of Al2CuMg and MgZn2 precipitates [24]. These precipitates were observed to intersect the dislocations during the hot tensile test deformation. These dislocation pile-ups cause the start of grain development at both the grain and grain borders, resulting in the creation of new recrystallized grains. Dynamic recrystallization causes the creation of serrated grain boundaries due to the softening behaviour that happens following the peak stress. The recrystallized grains formed at temperatures of 350, 475, and 525°C can still be attributed to the mechanisms of geometrical dynamic recrystallization (GDRX). The grain growth at higher temperatures results in coarser recrystallized grains as observed in Figure 5(c). DDRX has been previously seen under identical thermo-mechanical circumstances during the hot compression of aluminium alloys [25].

4. Conclusion

The study subjected the AA7075 alloy to various heat treatment conditions presented as R(T651), H (415°C-3 Hours-Air Cooled), H1 (350°C), H2 (400°C), H3 (450°C), H4 (475°C) H5 (500°C) and H6 (525°C). A study was done to investigate the hot ductility characteristics during a hot tensile test. Based on the examination of the acquired results, the following conclusions can be drawn.

1. The extension of the hot working range of aluminium alloy AA7075-T651 up to 525°C, instead of 477°C, can be inferred based on the absence of any notable properties or microstructural alterations. The yield and final stress at elevated temperatures significantly decrease from 445.25 MPa to 287.84 MPa and 580.33 MPa to 354.81 MPa respectively, which is necessary for processes involving forming, extruding, or other types of shape deformation.
2. The decrease in yield strength with increasing working temperature (350°c to 525°C), along with increased elongation (30% ductility at 525°C) and reduced necking area (40% on average for a temperature range of 450°C to 525°C), indicates that the material remains elastoplastic, which is advantageous for shaping or secondary processes.
3. The fracture mode can be determined by fractography, which reveals the presence of dimples deep shear dimples, and microvoids on the surface of the crack, confirming that the fracture mode is predominantly ductile for the H (350°C). The presence of dimples with cleavage facets exhibited ductile/ semi-ductile fracture mode for the H6 (525°C) condition. Semi-ductile fractures manifest when treated at temperatures exceeding the solidus temperature of 475°C.
4. The modulus of toughness of the H (350°C) to the H6 (525°C) was marginalized consistently between a workable range of 50-60 MPa. Thus, the hot working range for AA7075-T651 aluminium alloy can be extended to 525°C.

5. Dynamic recrystallization induces the formation of serrated grain boundaries as a result of the subsequent relaxation behaviour associated with the peak stress. The formation of recrystallized grains at temperatures of 350°C, 475°C, and 525°C can still be attributed to geometrical dynamic recrystallization (GDRX) mechanisms.
6. The evaluation of ductility required for different shaping procedures can be precisely determined by considering the operating temperature. The hot tensile test results indicated that the H4 (475°C) condition exhibited a ductility of 27%, whilst the H6 (525°C) condition had a maximum ductility of 30%.

The hot tensile test on AA7075-T651 at different temperatures showed that the material is suitable for secondary processes such as hot forging, hot rolling, and hot stamping. It also demonstrated an extended hot working capacity of up to 525°C, which is greater than the solidus temperature of 477°C. Post-treatment can be used to improve the strength of materials that may have been weakened during heat treatment or shape deformation procedures.

References

[1] A. Fakioglu, D. Özyürek, and R. Yilmaz, "Effects of different heat treatment conditions on fatigue behavior of AA7075 alloy," High Temp. Mater. Processes 32, 345–351 (2013).
[2] Aluminium Alloys - Design and Development of Innovative Alloys, Manufacturing Processes and Applications. 2022, doi: 10.5772/intechopen.95155
[3] Gu, R., Liu, Q., Chen, S. et al. Study on High-Temperature Mechanical Properties and Forming Limit Diagram of 7075 Aluminum Alloy Sheet in Hot Stamping. J. of Materi Eng and Perform 28, 7259–7272 (2019). https://doi.org/10.1007/s11665-019-04436-x.
[4] González-Velázquez, J. (2020). Plastic Deformation Mechanisms. In: Mechanical Behavior and Fracture of Engineering Materials. Structural Integrity, vol 12. Springer, Cham. https://doi.org/10.1007/978-3-030-29241-6_3.
[5] A. Asghari, A. Zarei-Hanzaki, M. Eskandari, Temperature dependence of plastic deformation mechanisms in a modified transformation-twinning induced plasticity steel, Materials Science and Engineering: A, Volume 579, 2013, Pages 150-156, ISSN 0921-5093, https://doi.org/10.1016/j.msea.2013.04.106.
[6] Tisza, Miklós & Budai, David & Kovács, Péter & Lukacs, Zsolt. Investigation of the formability of aluminium alloys at elevated temperatures. IOP Conference Series: Materials Science and Engineering. 159 2019, https://doi.org/10.1088/1757-899X/159/1/012012.
[7] B-A Behrens *et al* 2018 *IOP Conf. Ser.: Mater. Sci. Eng.* 418 012027, 10.1088/1757-899X/418/1/012027
[8] D.S., Gertsriken., A.M., Husak., V.F., Mazanko., S., Ye., Bogdanov. The deformation rate, atomic mobility and properties of metals. 97(1):28-37. doi: 10.15407/MOM2021.01.028
[9] Zhang, Xiao., Zheng, Ren., Xiu, Rong, Zhu., Ren, Jing., Fu, Kui, Lian., Li, Qun, Hou. (2014). Behavior of Flow Stress of Al-Mn Alloy during Hot Compression. Advanced Materials Research, 886:109-114. doi: 10.4028/WWW.SCIENTIFIC.NET/AMR.886.109
[10] H.J., McQueen., E., Evangelista. Hot Working Defines Thermomechanical Processing (TMP) for Aluminum and Composites. Materials Science Forum, 89-96, 2012. doi: 10.4028/www.scientific.net/MSF.706-709.89
[11] M.A., Jabbari, Taleghani., E.M., Ruiz, Navas., Maryam, Salehi., José, Manuel, Torralba. (2012). Hot deformation behaviour and flow stress prediction of 7075 aluminium alloy powder compacts during compression at elevated temperatures. Materials Science and Engineering A-structural Materials Properties Microstructure and Processing, 534:624-631. doi: 10.1016/J.MSEA.2011.12.019
[12] Anna, Bouzekova-Penkova., A., I., Miteva. Some Aerospace Applications of 7075 (B95) Aluminium Alloy. Aerospace Research in Bulgaria, 34:165-179, 2022. doi: 10.3897/arb.v34.e15
[13] A., Woźnicki., Beata, Leszczyńska-Madej., Grzegorz, Włoch., Justyna, Grzyb., J., Madura., D., Leśniak. Homogenization of 7075 and 7049 Aluminium Alloys Intended for Extrusion Welding. 11(2):338,2021. doi: 10.3390/MET11020338
[14] Olena, L., Skuibida. Aluminum as a promising material in the automotive industry. Grail of Science, 266-269, 2023. doi: 10.36074/grail-of-science.17.02.2023.049
[15] M, Jiménez-Martínez., Mariel, Alfaro-Ponce., Cristopher, Antonio, Muñoz-Ibáñez. Design of an Aluminum Alloy Using a Neural Network-Based Model. Superalloys, 12(10):1587-1587, 2022. doi: 10.3390/met12101587
[16] Patrick T Summers , Yanyun Chen, Christian M Rippe, Ben Allen, Adrian P. Mouritz , Scott W Case and Brian Y Lattimer, Overview of aluminum alloy mechanical properties during and after fires, Summers et al. Fire Science Reviews (2015) 4:3, DOI 10.1186/s40038-015-0007-5

[17] Fang, J., Zhu, Z., Zhang, X. et al. Tensile Deformation and Fracture Behavior of AA5052 Aluminum Alloy under Different Strain Rates. J. of Materi Eng and Perform 30, 9403–9411 (2021). https://doi.org/10.1007/s11665-021-06112-5

[18] Giuseppe, Mirone., Raffaele, Barbagallo. Modelling the influence of strain and strain rate on the thermal softening during dynamic loading of ductile metals. Epj Web of Conferences, 250:02003, 2021. doi: 10.1051/EPJCONF/202125002003

[19] Young, W., Kwon., K., S., Tan, Failure of Ductile Materials Subject to Varying Strain Rates. Journal of Pressure Technology-transactions of The Asme, 133(1):011402, 2011. doi: 10.1115/1.4002054

[20] T. Sunara, T. Tuncaya, D. Özyüreka, and M. Gürüb, Investigation of Mechanical Properties of AA7075 Alloys Aged by Various Heat Treatments, Physics of Metals and Metallography, 2020, Vol. 121, No. 14, pp. 1440–1446

[21] P. K. Rout, M. M. Ghosh, and K. S. Ghosh, "Microstructural, mechanical and electrochemical behaviour of a 7017 Al–Zn–Mg alloy of different tempers," Mater. Charact. 104, 49–60 (2015). 35.

[22] S. Malarvizhi, V. and Balasubramanian, "Effect of welding processes on AA2219 aluminium alloy joint properties," Trans. Nonferrous Met. Soc. China 21, 962–973 (2011).

[23] E. Acer, E. Çadirli, H. Erol, T. Kirindi, and M. Gündüz, "Effect of heat treatment on the microstructures and mechanical properties of Al–5.5Zn– 2.5Mg alloy," Mater. Sci. Eng., A 662, 144–156 (2016)

[24] M. Taheri-Mandarjani, A. Zarei-Hanzaki, H.R. Abedi, Hot ductility behavior of an extruded 7075 aluminum alloy, Materials Science & Engineering A, http://dx.doi.org/10.1016/j.msea.2015.03.038

[25] M.R. Rokni, A. Zarei-Hanzaki, H.R. Abedi, N. Haghdadi, Microstructure evolution and mechanical properties of backward thixoextruded 7075 aluminum alloy, Materials & Design (1980-2015), Volume 36, 2012, Pages 557-563, ISSN 0261-3069, https://doi.org/10.1016/j.matdes.2011.11.061.

Effect of Fibre Reinforcement on Compressive Strength and Compressive Modulus of Epoxy Granite

Deepak D Ubale*, Abhijeet P Shah

Department of Mechanical Engineering, RIT, Islampur, Maharashtra, India

*Email: ubaledd@gmail.com

Abstract. The main focus of the paper is to study the effect of fibre reinforcement on compressive strength and compressive modulus of epoxy granite. To investigate the effect of fibre reinforcement, two compositions were considered likes 20% Epoxy and 80% Granite without fibre reinforcement and 20% Epoxy, 75% Granite and 5% fibre reinforcement. Based on the literature studied, three types of fibre were considered viz. steel fibre, carbon fibre and basalt fibre. Three specimens of each type of fibre i.e. nine specimens and three specimens without fibre reinforcement were prepared. The obtained maximum compressive strength and compressive modulus of epoxy granite without reinforcement is 91.70 MPa and 3.5 GPa for the composition of 20% epoxy and 80% granite. The observed maximum compressive strength of 78.57 MPa for 20% epoxy and 75% granite with 5% carbon fibre reinforcement epoxy granite (CFEG) which is much higher than steel fibre reinforced epoxy granite (SFEG) with 20% epoxy and 75% granite and basalt fibre reinforced epoxy granite (BFEG) with 20% epoxy and 75% granite. The computed compressive modulus is 2.7 GPa for SFEG with 5% fibre which greater than CFEG and BFEG with the same fibre content.

Keywords: Epoxy Granite, Steel fibre, Carbon fibre, Basalt fibre, Compressive strength, Compressive modulus

1. Introduction

Inadequate stiffness of materials used for structural parts of machining centres is one of the causes of machining vibrations. The conventional materials used have high stiffness with inadequate damping for stated application. Therefore there is a need to have materials with sufficient stiffness and damping. These structural parts are produced by the conventional casting (Hot casting) process creates air pollution. Thus eco-friendly, inexpensive material is essential to produce such structural parts of machining centres. In this work, the waste granite pieces generated were used as a filler material with randomly oriented fibres as a reinforcement and epoxy resin as a matrix material. The main focus of this experimental investigation is to determine the effect of fibre contents on the compressive strength and compressive modulus of epoxy granite. The fibres used for this work are steel fibres, carbon fibres and basalt fibres. The obtained results are compared with epoxy granite

IOP Conf. Series: Materials Science and Engineering **1316** (2024) 012016 doi:10.1088/1757-899X/1316/1/012016

without fibre-reinforced. Such type of studies was not reported in the literature studied for this experimental work.

Antonio et al. [1] investigated the compression strength of PMC for different combinations of epoxy of 15% and 20% by weight and granite particles of 50% small and 50% large. It was observed that the maximum compression strength was 114.23 MPa for 20% epoxy and granite particles 50% small and 50% large. Antonio et al.[2] investigated the damping of epoxy granite for various compositions. 80% of granite and 20% of epoxy have shown damping nearly three times greater than grey cast iron. D'Mello et al. [3] performed the compression test on the samples of epoxy and granite particulate. The maximum compressive strength of 116 MPa was obtained for the composition of 20 % epoxy and 50-50 granite particulate ratio by weight. The maximum compressive strength was obtained as 110 MPa. Subrahmanya et al. [4] have tested the specimens for compression test. The specimen having 50% coarse aggregates offered the highest value of elastic modulus near 30 GPa and compressive strength of 125 MPa. S. S. Balakrishna et al. [5] conducted the compression test on epoxy granite filled with cast iron particles as filler in the ratio of 5, 10, and 15%. Maximum compressive strength was obtained as 112 MPa for 30% epoxy by weight. As cast iron filler content increases, the compressive strength also increases. Joel D'Mello et al. [6] obtained the maximum compressive strength of 111 MPa for the combination of 20% epoxy and 80% granite grit by weight. Mohamed Omar et al. [7] tested epoxy granite (Aswan red granite) for compressive strength. It was found that the highest compressive strength of 72.15 MPa for a composite with granite to epoxy ratio of 80:20% wt. Mohammed Y. Abdellah et al.[8] investigated the compressive strength of Epoxy Granite(EG) composite of epoxy resin by 12% weight and crushed granite particles 1.18–2.36 mm, 0.6–1.18 mm and ≤ 0.6 mm. The results obtained show that the composite material EG with fine granite particles has the highest compressive strength of 18.1 MPa. Aditya et al. [9] studied experimentally the damping behaviour of carbon fiber, basalt fiber and glass fiber-reinforced epoxy laminate beams. It was found that the glass fibre epoxy laminate showed the largest damping ratio followed by basalt fibre epoxy laminate and carbon fibre epoxy laminate. The maximum storage and loss modulus were observed for carbon fibre and basalt fibre respectively. Huseyin et al. [10] examine the effect of fibreglass on the tensile strength of epoxy and basalt granules with silica particles. The specimens with fibreglass reinforcement have shown ductile behaviour whereas a specimen without fibreglass reinforcement with high rigidity breaks without any elongation. Wenfeng Bai et al. [11] glass fibres were used to enhance the tensile strength and Young's modulus of polymer concrete. Glass fibre reinforced polymer concrete (GFRPC) is a mixture of glass fibre and polymer concrete. The damping tests were performed on GFRPC. It was found that an increase in epoxy resin increases the damping ratio of GFRPC. An increase in the content of glass fibre increases the damping ratio of GFRPC. The maximum value of the damping ratio was obtained when the glass fibre length was 20 mm.

2. Material and Method

2.1 Materials

In this study, three ingredients were used to prepare the test specimens namely Granite of Indian origin, Epoxy Resins, various fibres like steel, carbon and basalt. Araldite LY 556 of bisphenol A group as an epoxy and Ardur 951 as a hardner of Huntsman make are used. Fibres and epoxy resin were procured from Fibre Source, Chennai. The ratio of granite to epoxy was 80:20 by volume for test specimens without fibre reinforcement. The resin content was kept at 20% by volume for all compositions. The ratio of epoxy to hardener was 10:1 by weight. The tested Indian granite was provided by Accurate Gauging, Pune. The granite aggregates were sieved after the manual crushing of granite waste pieces. These aggregates were categorised as; <1 mm, 1- 3mm, 3-5 mm and 5-8 mm. The granite aggregate proportions considered are shown in table 1. In the fibre-reinforced test specimens, the ratio of granite to epoxy to fibre was kept at 75:20:5 by volume. The specifications of the fibres used are shown in Table 2.

Table 1. Granite Aggregate ratio

	<1 mm	1- 3mm,	3-5 mm	5-8 mm
Volume Percentage	20	30	30	20

Table 2. Fibres Specifications

Type	Steel fibre	Carbon fibre	Basalt fibre
Aspect Ratio	100	1700	600
Density(Kg/dm^3)	7.85	0.35	2.7

2.2 Specimen Preparation

Test specimens were prepared as per ASTM standard C39, C192/C. The cylindrical test specimens of a length-to-diameter ratio of nearly equal to 2 were prepared. The hollow cylindrical mould of diameter $75^{\pm 0.5}$ mm and length $150^{\pm 0.5}$ mm was manufactured which is shown in figure 1. The mixture of epoxy, granite aggregates and one type of fibre was stirred manually to achieve maximum homogeneity as possible. This mixture was poured into a mould cavity step by step on the vibration table. This filled mould was kept in natural conditions to cure for 24 hours at room temperature. A total of 12 specimens were prepared as mentioned in Table 3.

Figure 1. Hollow Cylindrical Mould

Table 3. Test Specimen Compositions (% by volume)

	Aggragate	Fibre	Resin	No. of Specimen
Without Fibre	80	0	20	3
Basalt Fibre	75	5	20	3
Steel Fibre	75	5	20	3
Carbon Fibre	75	5	20	3

IOP Conf. Series: Materials Science and Engineering **1316** (2024) 012016 doi:10.1088/1757-899X/1316/1/012016

2.3 Experimentation

The compression test was performed at the concrete laboratory, RIT Islampur on a Universal Testing Machine (UTM) of 1000 kN capacity. The test specimens were subjected to gradual compression load at a uniform rate without shock till the failure of the specimen. The readings of compressive load applied in Kilo-Newton against compression in millimetres were recorded in software interfaced with UTM. Stress-strain curves were generated in the same software. A total 12 number of compression tests were performed.

Figure 2. Compression Test

3. Results and Discussion

The aimed material characterization is a function of fibre type and its content considered. To evaluate the effect of fibre reinforcement, other parameters viz. ratio of epoxy and granite distribution were kept constant for fibre-reinforced compositions. The compression test specimens with fibres prepared contain one type of fibre at 5% and granite aggregates at 75% by volume. The epoxy of 20% by volume was used for both types of specimens i.e. with and without fibre reinforcement the obtained results of the compression test of the tested specimens are shown in Table 4.

IOP Conf. Series: Materials Science and Engineering **1316** (2024) 012016 doi:10.1088/1757-899X/1316/1/012016

Table 4. Compression Test Results

Composition	Specimen No.	Compressive Strength (MPa)
Granite: Epoxy(80:20)	EG1	87.59
	EG2	86.35
	EG3	91.70
Granite: Epoxy : Steel Fibres(75:20:5)	SFEG1	58.95
	SFEG2	49.35
	SFEG3	56.76
Granite: Epoxy :Carbon Fibres(75:20:5)	CFEG1	69.08
	CFEG2	78.57
	CFEG3	73.46
Granite: Epoxy :Basalt Fibres(75:20:5)	BFEG1	55.76
	BFEG2	48.74
	BFEG3	58.89

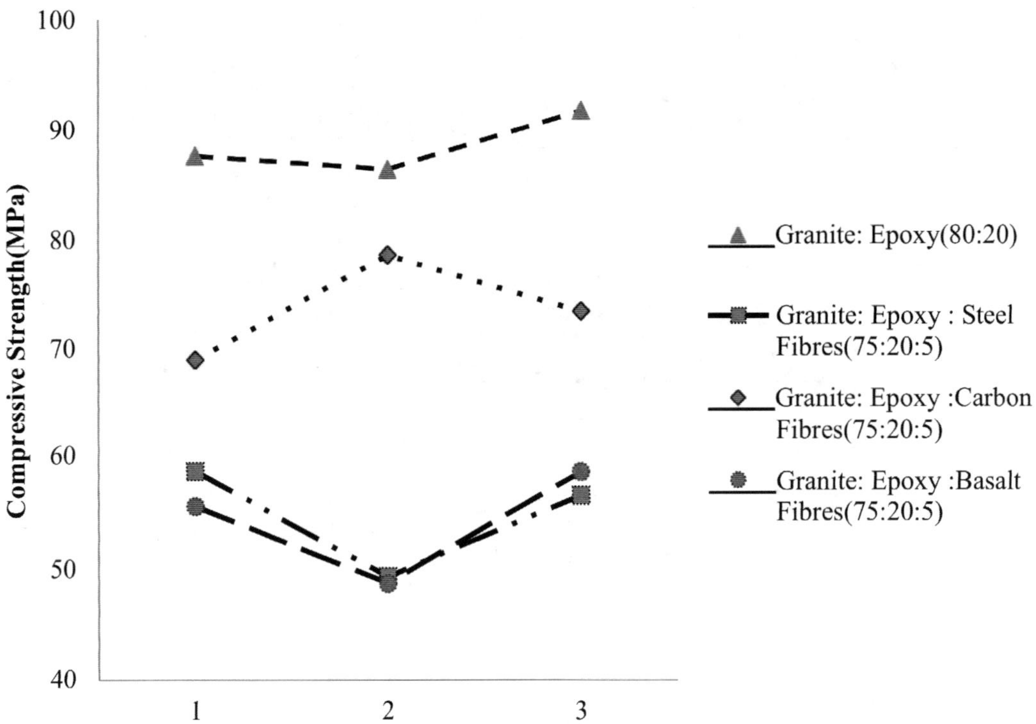

Figure 3. Compressive strgnth(MPa)

The variations of the obtained compressive strengths for various compositions tested are shown in figure 3. The maximum compressive strength was found 91.70 MPa for Granite: Epoxy (EG) of ratio 80:20 as compared to all other compositions. In fibre-reinforced compositions, carbon fibre-reinforced epoxy granite (CFEG) shows better results with a maximum compressive strength of 78.57

MPa. This value of compressive strength is much more than steel fibre-reinforced epoxy granite (SFEG) and basalt fibre-reinforced epoxy granite (BFEG). It was also observed that there is no significant difference in compressive strengths of SFEG and BFEG. Figure 4 shows the stress-strain curves for compositions studied for compression tests. The slope of the linear region of the stress-strain curve provides the compressive modulus (E_c) in GPa which is shown in Table 5. The computed E_c of epoxy granite without reinforcement is 3.5 GPa. For the fibre-reinforced compositions, the E_c is 2.7 GPa for SFEG followed by 2 GPa for CFEG then 1.5 GPa for BFEG.

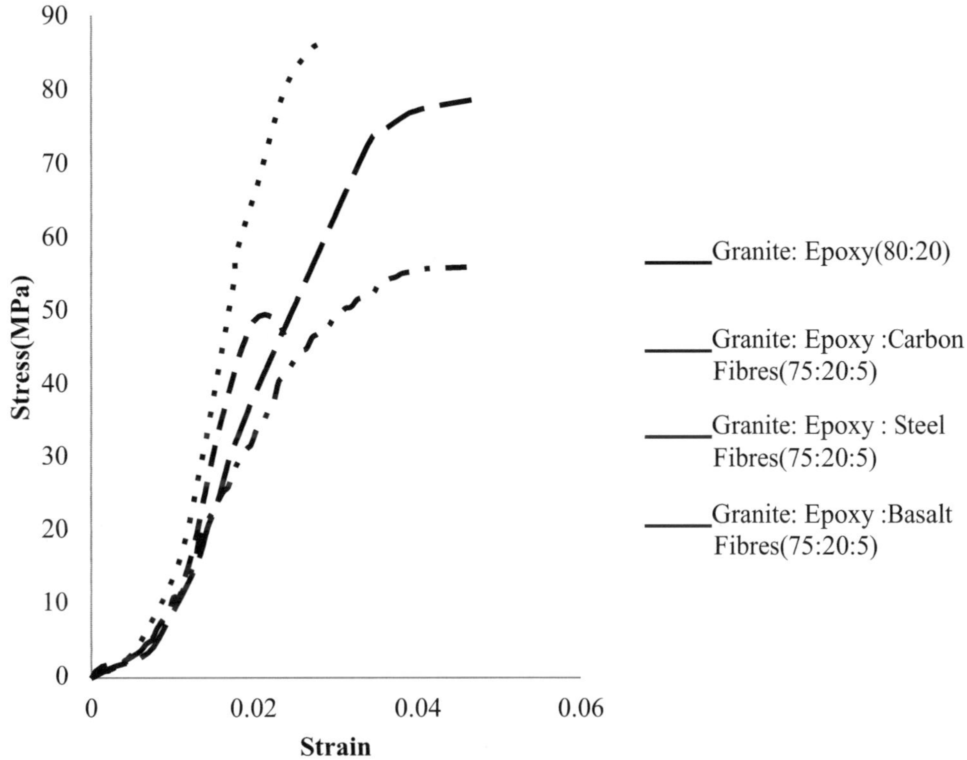

Figure 4. Stress-Strain Diagram

Table 5. Compressive Modulus of Epoxy Granite Compositions

Composition	Compressive Modulus (GPa)
Granite: Epoxy(80:20)	3.5
Granite: Epoxy : Steel Fibre(75:20:5)	2.7
Granite: Epoxy :Carbon Fibre(75:20:5)	2
Granite: Epoxy :Basalt Fibre(75:20:5)	1.5

IOP Conf. Series: Materials Science and Engineering **1316** (2024) 012016 doi:10.1088/1757-899X/1316/1/012016

EG

CFEG

SFEG

BFEG

Figure 5. Failure mode under axial compression loading

Figure 5 shows failure modes of tested specimens of various compositions under axial compression loading. In case of the EG specimens, side fracture is observed at top of the specimens. For CFEG specimens, partial shear failure is observed with little lateral strain. In case of SFEG and BFEG specimens, shear failure is observed.

4. Conclusion

In this work, the main aim was to investigate the effect of fibre reinforcement on the compressive strength and compressive modulus of epoxy granite and its comparison with epoxy granite without reinforcement. The maximum compressive strength and compressive modulus of epoxy granite without reinforcement is 91.70 MPa and 3.5GPa for 20% epoxy and 80% granite. 5% of carbon fibre reinforcement (CFEG) gives maximum compressive strength of 78.57 MPa with 20% epoxy and 75% granite which is much more as compared with SFEG and BFEG with 20% epoxy and 75% granite. The maximum compressive modulus obtained is 2.7 GPa for SFEG which is more as compared to CFEG and BFEG. Further increase in the fibre content may not improve the compressive strength and the compressive modulus of fibre-reinforced epoxy granite for compression loading. Fibre reinforcement may enhance the flexural strength of epoxy granite with some amount of breaking.

IOP Conf. Series: Materials Science and Engineering **1316** (2024) 012016 doi:10.1088/1757-899X/1316/1/012016

References

[1] Antonio Piratelli-Filho, Frank Shimabukuro, 2008, 'Characterization of Compression Strength of Granite-epoxy Composites Using Design of Experiments', Materials Research, Vol. 11, No. 4, pp.399-404.

[2] Antonio Piratelli Filho, Flaminio Levi Nito, 2010, 'Behavior of granite epoxy beams subjected to mechanical vibrations', Materials Research, 13(4), pp. 497-503.

[3] Joel Antony D'Mello, S. S. Balakrishna,2014, 'Assessment of Particulates Matrix Ratio and Its Influence on Mechanical Properties of Granite Particulates Reinforced Epoxy Composite', IJMET, Volume 5, Issue 9, September (2014), pp. 266-270

[4] Subrahmanya Swamy. S , B. R. Sreedhar , Vinayak J. Kalas, Chandan K. M,2014, 'Experimental Studies on Compression and Vibration Characteristics of Granite Epoxy- An Alternative Material for Precision Machine Tool Beds', IJPRET, 2014; Volume 2 (10): 120-135.

[5] S. S. Balakrishna, H. Girish, G. C. Mohan Kumar, S. Narendranath, 2016, 'Analysis on Mechanical and Dynamic behavior of Granite Epoxy Composites with Cast Iron Particulates as Filler', Indian Journal of Advances in Chemical Science S1 (2016) 122-126.

[6] Joel D'Mello, Alister G D'Souza, Sushanth H Gowda, and Denis Pinto,2019,' Experimental investigation of compression, flexural strength and damping behaviour of granite particulate epoxy matrix composite', AIP Conference Proceedings 2080, 020012 (2019).

[7] Mohamed Omar , Yasser Abdelrhman, Ibrahim M. Hassab-Allah , Wael M. Khierldeen,2021, 'Experimental Study on Compressive Strength and Flexural Rigidity of Epoxy Granite Composite Material', Journal of Engineering Sciences Assiut University Faculty of Engineering Vol. 49, No. 2 March 2021 PP. 198 – 214.

[8] Mohammed Y. Abdellah, Ahmed Abdelhaleem, Ibrahim A. Alnaser, G. T. Abdel-Jaber andAbdalla Abdal-hay,2021, 'Flexural, Compression and Fracture Properties of Epoxy Granite as a Cost-Effective Structure Materials: New Machine Element Foundation', AIMS Materials Science, 8(1): 82–98.

[9] Aditya Raj,Jayant Prakash Varun, P. K. Mahato,2019, 'Fabrication and Vibration Damping Analysis of Basalt Fiber Reinforced Composite Beam', AIP Conference Proceedings 2134, 080002 (2019).

[10] Huseyin Onur Ozturk, Yaşar Kahraman, 2019, 'Effects of glass fiber Reinforcement to Tensile Strength in Epoxy Matrix Granular Composite Materials' Sakarya University Journal of Science 23(5), 736-743.

[11] Wenfeng Bai , Jianhua Zhang , Peng Yan , Xinli Wang , 2008, 'Study on vibration alleviating properties of glass fiber reinforced polymer concrete through orthogonal tests', Materials & Design, 30(4), 1417–1421.

AUTHOR INDEX